Manuela Winkler, Karl Hülber, Leticia Cruz-Paredes,
José García-Franco, Klaus Mehltreter,
Angélica Jímenez-Aguilar und Peter Hietz

Population Dynamics of Epiphytes

Manuela Winkler, Karl Hülber,
Leticia Cruz-Paredes, José García-Franco,
Klaus Mehltreter, Angélica Jímenez-Aguilar
und Peter Hietz

Population Dynamics of Epiphytes

The Influence of Host Branches

Südwestdeutscher Verlag für Hochschulschriften

Impressum/Imprint (nur für Deutschland/ only for Germany)
Bibliografische Information der Deutschen Nationalbibliothek: Die Deutsche Nationalbibliothek verzeichnet diese Publikation in der Deutschen Nationalbibliografie; detaillierte bibliografische Daten sind im Internet über http://dnb.d-nb.de abrufbar.

Alle in diesem Buch genannten Marken und Produktnamen unterliegen warenzeichen-, marken- oder patentrechtlichem Schutz bzw. sind Warenzeichen oder eingetragene Warenzeichen der jeweiligen Inhaber. Die Wiedergabe von Marken, Produktnamen, Gebrauchsnamen, Handelsnamen, Warenbezeichnungen u.s.w. in diesem Werk berechtigt auch ohne besondere Kennzeichnung nicht zu der Annahme, dass solche Namen im Sinne der Warenzeichen- und Markenschutzgesetzgebung als frei zu betrachten wären und daher von jedermann benutzt werden dürften.

Verlag: Südwestdeutscher Verlag für Hochschulschriften Aktiengesellschaft & Co. KG
Dudweiler Landstr. 99, 66123 Saarbrücken, Deutschland
Telefon +49 681 37 20 271-1, Telefax +49 681 37 20 271-0, Email: info@svh-verlag.de
Zugl.: Wien, Universität für Bodenkultur, Dissertation, 2005

Herstellung in Deutschland:
Schaltungsdienst Lange o.H.G., Berlin
Books on Demand GmbH, Norderstedt
Reha GmbH, Saarbrücken
Amazon Distribution GmbH, Leipzig
ISBN: 978-3-8381-0281-8

Imprint (only for USA, GB)
Bibliographic information published by the Deutsche Nationalbibliothek: The Deutsche Nationalbibliothek lists this publication in the Deutsche Nationalbibliografie; detailed bibliographic data are available in the Internet at http://dnb.d-nb.de.

Any brand names and product names mentioned in this book are subject to trademark, brand or patent protection and are trademarks or registered trademarks of their respective holders. The use of brand names, product names, common names, trade names, product descriptions etc. even without a particular marking in this works is in no way to be construed to mean that such names may be regarded as unrestricted in respect of trademark and brand protection legislation and could thus be used by anyone.

Publisher:
Südwestdeutscher Verlag für Hochschulschriften Aktiengesellschaft & Co. KG
Dudweiler Landstr. 99, 66123 Saarbrücken, Germany
Phone +49 681 37 20 271-1, Fax +49 681 37 20 271-0, Email: info@svh-verlag.de

Copyright © 2009 by the author and Südwestdeutscher Verlag für Hochschulschriften Aktiengesellschaft & Co. KG and licensors
All rights reserved. Saarbrücken 2009

Printed in the U.S.A.
Printed in the U.K. by (see last page)
ISBN: 978-3-8381-0281-8

Table of contents

1. **General Introduction** 7
   1.1 Epiphyte demography 7
   1.2 Study site 9
   1.3 Study species 10
   1.4 Objectives 11
   1.5 Literature 15
2. **Herbivory in epiphytic bromeliads, orchids and ferns in a Mexican montane forest.** 17
   2.1 Introduction 18
   2.2 Methods 19
       Study site 19
       Study species 19
       Folivory 20
       Herbivory in reproductive organs 21
       Herbivory in bromeliad ramets 21
       Statistical analysis 22
   2.3 Results 22
       Folivory 22
       Herbivory in reproductive organs 26
       Herbivory in bromeliad ramets 26
   2.4 Discussion 26
       Folivory 26
       Herbivory in reproductive organs 29
       Herbivory in bromeliad ramets 29
       Conclusions 30
   2.5 Acknowledgements 30
   2.6 Literature 31
3. **Effect of canopy position on germination and seedling survival of epiphytic bromeliads in a Mexican humid montane forest** 34
   3.1 Introduction 35
   3.2 Materials and Methods 36
       Study area and species 36
       Germination experiments 37
       Census of natural populations 38
       Statistical methods 38
   3.3 Results 40
       Germination experiments 40
       Census of natural populations 43
   3.4 Discussion 45
       Germination experiments 45
       Census of natural population 47
       Conclusions 48
   3.5 Acknowledgements 48
   3.6 Literature 49
4. **Breeding systems, fruit set, and flowering phenology of epiphytic bromeliads and orchids in a Mexican humid montane forest** 52
   4.1 Introduction 53
   4.2 Methods 55
   4.3 Results 57
   4.4 Discussion 60
   4.5 Acknowledgements 64

1

    4.6   Literature 65
5   Spatial and temporal variability in the population dynamics of epiphytic orchids in a Mexican humid montane forest 68
    5.1   Introduction 69
    5.2   Material and Methods 70
        Study area and species 70
        Population census 70
        Matrix population models 71
        Logistic Regression Models 73
    5.3   Results 75
        Demographic properties 75
        Elasticity analysis 79
        Influence of branch characteristics 79
    5.4   Discussion 80
        Demographic properties 80
        Elasticity analysis 83
        Influence of branch characteristics 84
        Conclusions 84
    5.5   Acknowledgements 84
    5.6   Literature 85
    5.7   Appendix 88
6   Population dynamics of epiphytic bromeliads: Life strategies and the role of host branches 91
    6.1   Introduction 92
    6.2   Material and Methods 93
        Study area and species 93
        Population census 94
        Matrix population models 95
        Logistic regression models 97
    6.3   Results 97
        Demographic properties 97
        Elasticity analysis 99
        Effects of disturbance and microsite characteristics 101
    6.4   Discussion 103
        Demographic patterns 103
        Vital rates determining population dynamics 105
        Effects of disturbances and microsite characteristics 106
        Conclusions 107
    6.5   Acknowledgements 108
    6.6   Literature 109

List of Tables

Table 2.1	Leaf area lost assigned to folivory	24
Table 2.2	Percentage of individuals with and without folivory	24
Table 2.3	Foliar nitrogen concentration and leaf life span	25
Table 3.1	Percentage of seed germination and seedling survival	40
Table 3.2	Parametric survival models	44
Table 3.3	Median survival time in experimental and census populations	42
Table 3.4	Per cent mortality per day on shady and sunny branches	42
Table 4.1	Average number of flowers	57
Table 4.2	Results of pollination experiments	59
Table 4.3	Natural fruit set in bromeliads	60
Table 5.1	Criteria to distinguish stage classes	74
Table 5.2	Pooled transition matrices and demographic properties of populations	76
Table 5.3	Matrix properties	76
Table 5.4	Influence of branch parameters on survival	81
Table 5.5	Influence of branch parameters on growth	81
Table 5.6	Single year transition matrices and demographic properties	88
Table 6.1	Definition of stage classes	95
Table 6.2	Pooled transition matrices and demographic properties of populations	98
Table 6.3	Population growth rates with and without branchfall	99
Table 6.4	Influence of branch parameters on probability to become reproductive	107

List of Figures

Figure 1.1	Climatograph of Xalapa	10
Figure 1.2	Humid montane forest at the study site	12
Figure 1.3	Study species	13
Figure 1.4	Relative position of census trees	12
Figure 2.1	Folivory at the beginning of the dry season 2003	23
Figure 2.2	Leaf nitrogen content and mean leaf area consumed	23
Figure 2.3	Herbivory in bromeliad inflorescences	27
Figure 2.4	Infested bromeliad ramets	27
Figure 3.1	Kaplan-Meier estimates of seedling survival of bromeliad species	41
Figure 3.2	Effect of branch parameters on seedling survival	42
Figure 3.3	Mortality per day in different seasons	43
Figure 3.4	Kaplan-Meier estimates of seedling survival in experiment and census	45
Figure 4.1	Rainfall and flowering season	58
Figure 4.2	Fruit set of orchids during three years	60
Figure 5.1	Life cycle graphs	72
Figure 5.2	Elasticities of lambda of epiphytic orchids	77
Figure 5.3	Added elasticity values by stage and vital rate of epiphytic orchids	78
Figure 5.4	Population growth rate with and without branchfall	79
Figure 6.1	Elasticities of lambda of epiphytic bromeliads	100
Figure 6.2	Added elasticity for demographic processes and stages of bromeliads	101
Figure 6.3	Branch diameter and demographic processes	102
Figure 6.4	Survival probabilities depending on branch parameters	103

Abstract

Epiphytes are a characteristic and fascinating life-form of the humid tropics. They account for eight percent of global vascular plant diversity and harbour many animal species. However, little is known about their population dynamics and the influence of the host tree and branch characteristics on their life-cycles. Epiphytes need to balance the chances and risks of long-term survival and those of colonizing new branches in their high-stress and high-risk canopy habitats.

Populations of three orchid and five bromeliad species were censused over three and two years, respectively. Additionally, germination and pollination experiments were conducted. Matrix population models showed that average population growth rates were below unity in all species except for the xeric Tillandsia juncea, and in two other species in one of the years studied. Experimental germination rates in bromeliads ranged from 7.2 to 33.7 %. Subsequent seedling survival was low, which was also reflected in the stage distribution with juvenile stages dominating most species. However, seedling survival had only little effects on population growth rates, which were most sensitive to the survival of large pre-reproductive or reproductive stages. Mortality mostly decreased steadily with age.

The time to reach maturity ranged from six to 16 years. Canopy position and microsite characteristics influenced time to germination, survival rates and the probability of growing from the pre-reproductive to the reproductive stage. In most species these vital rates increased with the amount of available light. The breeding systems of the epiphytes investigated ranged from dioecious, to largely or entirely auto-incompatible and outcrossing to partly or mainly self-pollinating. Xeric species tented to be self-pollinating. Species growing on small and less stable branches had a higher fruit set through auto-pollination or other strategies, which may be of importance to compensate the high disturbance on these substrates. However, the influence of fecundity on population growth rates was small in all species.

Herbivory in reproductive organs was substantial in the orchid Lycaste aromatica and in several bromeliad species, but had a negligible effect on population performance. Folivory was less than 1.5 % in bromeliads and orchids, but much higher (7-20%) in five epiphytic fern species. Damage was positively correlated with leaf nitrogen content but not with leaf life-span. Many bromeliads were infested by curculionid larvae feeding on the meristematic tissue at the ramet base which accounts for a substantial percentage of ramet and shoot death of large individuals.

Most of the populations studied appear to be declining. Combined with the increasing loss of natural habitats this may lead to the extinction of species, especially those restricted to closed forests.

1 General Introduction

Manuela Winkler

1.1 Epiphyte demography

Vascular epiphytes are a characteristic and fascinating life-form of the humid tropics. Epiphytes make up 8 % of the total vascular plant diversity (Benzing 1990). In one of the most diverse tropical lowland rainforests they even account for a third of the plant diversity (Gentry & Dodson 1987). Besides the lack of root contact with the soil, epiphytes differ in two important features from terrestrial plants. They are distributed in a three-dimensional space clearly defined by the host tree's architecture, and their population dynamics is superimposed on the dynamics of the supporting branches and trees. As small branches break more frequently epiphytes on these face a higher risk of falling to the ground and dying (Hietz 1997). Furthermore, there are strong microclimatic and resource gradients related to the structure of the canopy, above all light and humidity. However, canopy soils, wind (through increased evaporation and the higher risk of falling from or with the host branch), and temperature may also play a vital role (Parker 1995).

The differential distribution of epiphytes within the canopy is well established (e. g., Freiberg 1996, Hietz & Hietz-Seifert 1995b, Johansson 1974, Kelly 1985, Merwin *et al.* 2003, Oliver 1930, Zotz 1997). Ecophysiological adaptations of epiphytes (reviewed in Lüttge 1989, Zotz & Hietz 2001) may in part explain this distribution, but many other factors determine seed dispersal, establishment, growth and reproduction of epiphytes. Little is known about the population dynamics of epiphytes, and even less about the interactions between microsites and ecological and demographic traits of epiphytes.

Most epiphytes disperse through anemochorous diaspores (Madison 1977) although their probability of reaching a suitable germination site should be substantially lower than for animal-dispersed seeds. However, many epiphytes may not be able to afford the additional cost of producing attractive diaspores (Benzing 1990). Anemochorous seeds of epiphytes tend to be small and numerous to compensate for the small chance of landing on a safe site (Madison 1977). The extreme case are orchids: their minute seeds are so reduced that they have to find a suitable micorrhizal fungus for germination, in addition to a branch. Observing the dispersal of epiphytic diaspores is even more difficult than for terrestrial plants. Experiments (Ackerman et al. 1996, García-Franco & Rico-Gray 1988, Mondragón 2000) and ballistic model calculations (Murren & Ellison 1998) suggest potentially high diaspore mobility depending on turbulence and height of release. However, most seeds remained near the dispersing mother "plant" or travelled only a few meters (García-Franco & Rico-Gray 1988, Mondragón 2000). Germination and subsequent seedling survival tended to be low in the field (e. g., Benzing 1978, 1981, Hernández-Apolinar 1992, Larson 1992, Mondragón 2000, Zotz 1998).

Survival probabilities of epiphytes on the forest floor are very low (Matelson et al. 1993), thus branchfall constitutes an important factor in epiphyte mortality (Hietz 1997). Excluding branchfall, mortality decreased with plant size (Hietz 1997, Zotz 1998). Most juvenile mortality occurred in the dry season suggesting that water is limiting juvenile survival (Benzing 1978, 1981, Larson 1992, Zotz 1998). This is also supported by physiological observations (Zotz & Ziegler 1999). Considering the existing microclimatic gradients within the canopy, an effect of plant position on mortality is to be expected.

Growth is generally slow in epiphytes, as a result of their resource-limited habitat (Benzing 1990). The species investigated so far are estimated to reach maturity between 6 and 21 years (Benzing 1981, 1990, Hernández-Apolinar 1992, Hietz et al. 2002, Larson 1992, Zotz 1995, 1998, Winkler & Hietz 2001), with the exception of twig epiphytes (Chase 1987). Bromeliads preferentially dwelling on thinner branches mature faster (Hietz et al. 2002).

Many epiphytes, especially orchids, have developed efficient pollination systems to ensure pollen transport despite their patchy and often dispersed distribution (Ackerman 1986), whereas anemogamy is virtually absent (Benzing 1990). Other strategies including self-compatibility, cleistogamy and autogamy are common among epiphytes (Catling 1990, Dressler 1981), above all in unfavourable, xeric environments (Benzing 1978, Soltis et al. 1987). Pollen mobility is an important factor determining population size, gene-flow between subpopulations and the degree of inbreeding.

Small size- or age-classes dominated the population structure of *Dimerandra emarginata* (Zotz 1998), *Encyclia tampensis* (Larson 1992), *Tillandsia utriculata* (Bennett 1991, calculated to homogeneous size-classes), *Laelia speciosa* (Hernández 1992 except in a population with a high frequency of flower collection), indicating high juvenile mortality. Small stages also predominated in *Jacquiniella leucomelana*, a species on small and mostly exposed branches, whereas size-classes were evenly distributed in *Lycaste aromatica*, a plant from large and mostly shady branches (Winkler & Hietz 2001).

Only a few attempts have been made to relate differences in population structure within a species to the distribution on the host trees. Zotz (1997) found no effect of branch diameter on the distribution for juveniles and adults of three bromeliad species. Though survival was higher in the tree periphery and stem base than on intermediate branches, and on the sides rather than at the top or underside of branches in exposed seedlings of these bromeliads, this site-specific survival was not related to and could not help to explain the distribution of older individuals. (Zotz and Vollrath, 2002). Hietz (1997) reported a higher proportion of juveniles on thin branches. In *Jacquiniella leucomelana*, an orchid mostly found on thinner and exposed branches, the average plant size and the proportion of fertile plants increased with branch size, but in *Lycaste aromatica*, large plants mostly growing on thick branches, the average plant size decreased with branch size and fertility increased with branch height (Winkler & Hietz 2001). This suggests that branch stability was limiting for *J. leucomelana* but light (related to branch height) for *L. aromatica*. Only a few population models have been developed for epiphytes (Calvo 1993, Calvo & Horvitz 1990, Hernández 1992, Mondragón *et al.* 2004, Tremblay 1998, Tremblay & Ackerman 2001), none did account for tree-epiphyte relationships.

1.2 Study site

All field experiments and the population census were conducted in the Botanical Garden Clavijero and a small forest reserve (ca. 29 ha) adjacent to the Instituto de Ecología ('Parque Ecológico Francisco Javier Clavijero' or 'Santuario Bosque de Niebla'), 2.5 km south of Xalapa, in central Veracruz, Mexico (19°31'N, 96°57'W), at 1350 m elevation. Mists are frequent especially in the winter months. Average annual temperature is 19° C, and annual precipitation is 1500 mm, most of which falls in the June-October wet season (Figure 1.1). The core of the reserve is undisturbed old-growth forest. In parts of the reserve extensive planting of coffee and citrics took place but it left most large forest trees intact and was abandoned several decades ago (Hietz 1997).

According to the Holdridge life-zone system (Holdridge 1967), the forest is at the transition between premontane and lower montane moist forest. In Mexico, it is commonly classified as 'bosque mesófilo de montaña' (mesophilous montane forest) according to Rzedowski (1986). Canopy height was ca. 24 m in plots studied by Williams-Linera (1997). Dominant tree species are *Quercus germana*, *Qu. xalapensis*, *Liquidambar macrophylla* and *Carpinus caroliniana*. These species are of temperate origin and deciduous, in contrast to trees of lower strata and shrubs, which are evergreen tropical elements (*Chamaecrista chamaecristoides*, *Chamaedorea spp.*, *Cuphea nitidula*, *Desmodium spp.*, *Hoffmannia excelsa*, *Oreopanax xalapensis*, *O. captitatus*, *Xylosma flexuosum*, Figure 1.2). Detailed descriptions of the forest structure are given by Williams-Linera (1997). Epiphytes, especially bromeliads (*Catopsis sessiliflora*, *Tillandsia butzii*, *T. deppeana*, *T. juncea*, *T.multicaulis*, *T. punctulata*) ferns (*Phlebodium areolatum*, *Pleopeltis crassinervata*, *Polypodium* spp.) and orchids (*Dichaea neglecta*, *Encyclia ochracea*, *Gongora galeata*, *Jacquiniella leucomelana*, *J. teretifolia*, *Scaphyglottis livida*), are very abundant. In the reserve 68 species of vascular epiphytes have been found (P. Hietz, pers. comm.). The epiphyte community is described in more detail by Hietz & Hietz-Seifert (1995a).

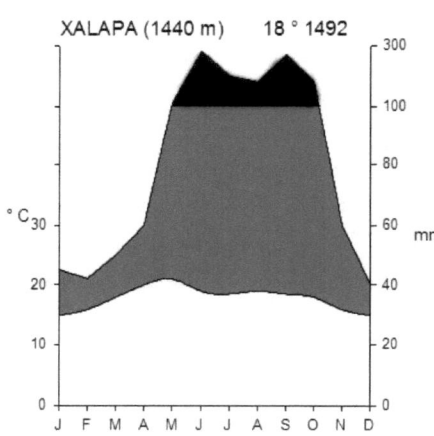

Figure 1.1: Climatograph of Xalapa, Veracruz (after Castillo-Campos 1991)

1.3 Study species

Studied species were the orchids *Jacquiniella leucomelana* (Reichenbach f.) Schlechter, *J. teretifolia* (Sw.) Britton & P. Wilson and *Lycaste aromatica* (Graham ex Hook.) Lindley) and the bromeliads *Catopsis sessiliflora* (Ruiz & Pav.) Mez, *Tillandsia deppeana* Steud., *T. juncea* (Ruiz & Pav.) Poir., *T. multicaulis* Steud. and *T. punctulata* Schltdl. & Cham.). *Jacquiniella teretifolia* has caespitose erect stems with distichous, almost terete leaves and yellowish-green flowers. *Jacquiniella leucomelana* is similar but stems, leaves and flowers are much smaller. *Lycaste aromatica* has thin, drought-deciduous leaves at the top of a broad pseudobulb. Dark-yellow flowers arise from the base of the pseudobulb. They emit a strong smell of cinnamon and are pollinated by euglossine bees (Dressler 1968). Monocarpic *T. deppeana* has a large impounding rosette and a tall, pinnate inflorescence tall, with reddish bracts and blue to violet corollas. *Tillandsia multicaulis* has also bright reddish bracts and blue corollas, but a smaller rosette and several sessile spikes, not exceeding the leaves. *Tillandsia punctulata* is of tank-atmospheric intermediate habit, the inflorescence is about as long or slightly longer than the leaves and composed of few, densely digitate spikes with red bracts and dark violet petals with a white apex. The atmospheric *T. juncea* has filiform, fasciculate leaves, the inflorescence is rather small, composed of a few, dense spikes, the bracts are reddish but less conspicuous than in the congeners studied and the petals are violet. In contrast to the other species, which are hermaphroditic, *C. sessiliflora* is dioecious, with the leaves forming small and narrow tanks, the inflorescence being pinnate or bi-pinnate with creamish flowers and very inconspicuous, green bracts (Hietz & Hietz-Seifert 1994, Figure 1.3).

1.4 Objectives

The main objective of this thesis is to identify demographic and ecological pattern of a set of abundant epiphytic orchids and bromeliads and to relate these features to the epiphytes' position on the tree and to key characteristics expected to influence epiphyte population dynamics: branch diameter (determining branch stability), light, atmospheric humidity, and substrate (bark, lichens, bryophytes, detritus). This should lead to a better understanding of epiphyte population dynamics as affected by their host trees and, as a more applied aspect, to understand what affects the potential of epiphytes to colonize new habitats.

Sixteen trees were sampled in the reserve to study epiphyte populations (Figure 1.4) between August 2001 and February 2005. For bromeliads, 186 branch sections on nine trees were delimited and all individuals of the study species on these sections were tagged. All orchids on each of six trees were sampled, with the exception of inaccessible individuals. Characteristics of sampled branches (height, distance to trunk, diameter, canopy openness, bryophyte and lichen cover, percentage of bare bark surface) were recorded. In addition to the population censuses, germination and pollination experiments were conducted in 2002/03 and herbivory in the study species and the most abundant epiphytic fern species in this forest was studied.

Figure 1.2: 'Bosque mesófilo de montaña' (Mesophilous montane forest) at the study site.

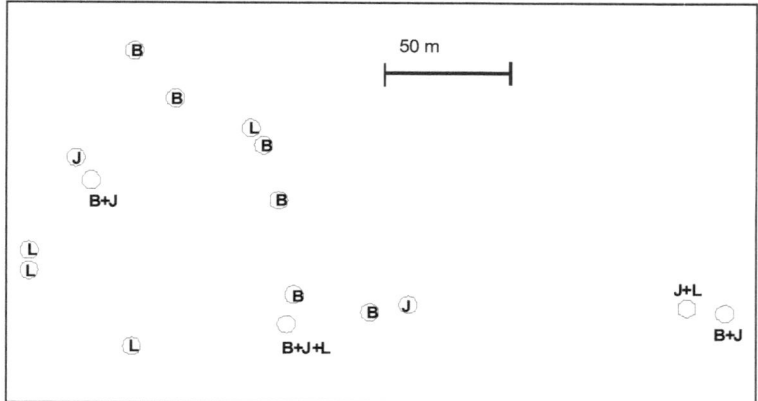

Figure 1.4: Relative position of trees with census populations. B = with bromeliads, J = with *Jacquiniella* spp. and L = with *Lycaste aromatica* censused.

A) *Jacquiniella leucomelana* (Photo: M.W.)

B) *Jacquinella teretifolia* (Photo: M.W.)

C) *Lycaste aromatica* (Photo: M.W.)

D) *Catopsis sessiliflora* (Photo: P.H.)

Figure 1.3: Study species.

E) *Tillandsia deppeana* (Photo: P.H.)

F) *Tillandsia juncea* (Photo: P.H.)

G) *Tillandsia multicaulis* (Photo: P.H.)

H) *Tillandsia punctulata* (Photo: P.H.)

This book covers a wide range of ecological topics related to the population dynamics of eight of the most important epiphytic species in mid-elevations of Veracruz. In Chapter 2, we present a study on epiphyte herbivory. Germination and seedling establishment of epiphytic bromeliads are analysed in detail in Chapter 3. Pollination, breeding system and flowering phenology are treated in Chapter 4. Finally, population models synthesize these findings and set them into a common context (Chapters 5 and 6).

1.5 Literature

Ackerman J. D. 1986. Coping with the epiphytic existence. Pollination strategies. *Selbyana* **9**: 52-60.

Ackerman J. D., Sabat A. & Zimmerman J. K. 1996. Seedling establishment in an epiphytic orchid: an experimental study of seed limitation. *Oecologia* **106**:192-198.

Bennett B. C. 1991. Comparative biology of Neotropical epiphytic and saxicolous *Tillandsia* spp.: Population structure. *Journal of Tropical Ecology* **7**: 361-371.

Benzing D. H. 1978. Germination and early establishment of *Tillandsia circinnata* Schlecht. (Bromeliaceae) on some of its hosts and other supports in southern Florida. *Selbyana* **5**: 95-106.

Benzing D. H. 1981. The population dynamics of *Tillandsia circinnata* (Bromeliaceae): cypress crown colonies in southern Florida. *Selbyana* **5**: 256-263.

Benzing D. H. 1990. *Vascular epiphytes*. Cambridge: Cambridge University Press.

Calvo R. N. & Horvitz C. C. 1990. Pollinator limitation, cost of reproduction, and fitness in plants: a transition-matrix demographic approach. *American Naturalist* **136**: 499-516.

Calvo R. N. 1993. Evolutionary demography of orchids: intensity and frequency of pollination and the cost of fruiting. *Ecology* **74**: 1033-1042.

Castillo-Campos G. 1991. *Vegetacion y Flora del Municipio de Xalapa, Veracruz*. Programa del Hombre y la Biosfera (MAB, UNESCO). Xalapa, México: Instituto de Ecología, A.C.

Catling P. M. 1990. Auto-pollination in the Orchidaceae. In: Arditti J. (ed.). *Orchid Biology, Reviews and Perspectives V*. Portland, Oregon: Timber Press. Pp. 121-158.

Dressler R. L. 1981. *The Orchids*. Cambridge, MA: Harvard Univ. Press.

Freiberg M. 1996. Spatial distribution of vascular epiphytes on three emergent canopy trees in French Guiana. *Biotropica* **28**: 345-355.

García-Franco J. G. & Rico-Gray V. 1988. Experiments on seed dispersal and deposition patterns of epiphytes. The case of *Tillandsia deppeana* Steudel (Bromeliaceae). *Phytologia* **65**: 73-78.

Gentry A. H. & Dodson C. 1987. Contribution of nontrees to species richness of a tropical rain forest. *Biotropica* **19**: 149-156.

Hernández-Apolinar M. 1992. *Dinámica poblacional de* Laelia speciosa *(H.B.K.) Schltr. (Orchidaceae)*. Tesis de Licenciatura, UNAM, Mexico, D.F.

Hietz P. & Hietz-Seifert U. 1994. *Epiphytes of Veracruz: An illustrated guide for the regions of Xalapa and Los Tuxtlas, Veracruz*. Xalapa: Instituto de Ecología.

Hietz P. & Hietz-Seifert U. 1995a. Intra- and interspecific relations within an epiphyte community in a Mexican humid montane forest. *Selbyana* **16**: 135-140.

Hietz P. & Hietz-Seifert U. 1995b. Structure and ecology of epiphyte communities of a cloud forest in central Veracruz, Mexico. *Journal of Vegetation Science* **6**: 719-728.

Hietz P. 1997. Population dynamics and disturbance of epiphytes in a Mexican humid montane forest. *Journal of Ecology* **85**: 767-775.

Hietz P., Ausserer J. & Schindler G. 2002. Growth, maturation and survival of epiphytic bromeliads in a Mexican humid montane forest. *Journal of Tropical Ecology* **18**: 177-191.

Holdridge L. R. 1967. *Life zone ecology*. San José, Costa Rica: Tropical Science Center.

Johansson D. 1974. Ecology of vascular epiphytes in West African rain forest. *Acta Phytogeographica Suecica* **59**: 1-129.

Kelly D. L. 1985. Epiphytes and climbers of a Jamaican rain forest: vertical distribution, life forms and life history. *Journal of Biogeography* **12**: 223-241.

Larson R. J. 1992. Population dynamics of *Encyclia tampensis* in Florida. *Selbyana* **13**: 50-56.

Lüttge U. 1989. *Vascular plants as epiphytes: evolution and ecophysiology*. Heidelberg: Springer Verlag.

Madison M. 1977. Vascular epiphytes: their systematic occurrence and salient features -. *Selbyana* **2**: 1-13.

Matelson T. J., Nadkarni N. M. & Longino J. T. 1993. Longevity of fallen epiphytes in a neotropical montane forest. *Ecology* **74**: 265-269.

Merwin M. C., Rentmeester S. A. & Nadkarni N. M. 2003. The influence of host tree species on the distribution of epiphytic bromeliads in experimental monospecific plantations, La Selva, Costa Rica. *Biotropica* **35**: 37-47.

Mondragón D., Durán R., Ramírez I. & Valverde T. 2004. Temporal variation in the demography of the clonal epiphyte *Tillandsia brachycaulos* (Bromeliaceae) in the Yucatán Peninsula, Mexico. *Journal of Tropical Ecology* **20**: 189-200.

Mondragón D. 2000. *Dinámica poblacional de Tillandsia brachycaulos Schltdl. en el parque nacional de Dzibilchaltún, Yuc.* PhD thesis. Centro de Investigación Científica de Yucatán, México.

Murren C. J. & Ellison. A. M. 1998. Seed dispersal characteristics of *Brassavola nodosa* (Orchidaceae). *American Journal of Botany* **85**: 675-680.

Oliver W. R. 1930. New Zealand epiphytes. *Journal of Ecology* **18**: 1-50.

Parker G. G. 1995. Structure and microclimate of forest canopies. In: Lowman M. E. & Nadkarni N. M. (eds). *Forest canopies*. Academic Press, San Diego. Pp 73 -106.

Rzedowski J. 1986. *Vegetación de México, 3rd edition*. Mexico: Editorial Limusa.

Soltis D. E., Gilmartin A. J., Rieseberg L. & Gardner S. 1987. Genetic variation in the epiphytes *Tillandsia ionantha* and *T. recurvata* (Bromeliaceae). *American Journal of Botany* **74**: 531-537.

Tremblay R. L. 1998. *Lepanthes caritensis*, an endangered orchid: no sex, no future? *Selbyana* **18**: 160-166.

Tremblay R. L. & Ackerman J. D. 2001. Gene flow and effective population size in *Lepanthes* (Orchidaceae): a case for genetic drift. *Biological Journal of the Linnean Society* **72**: 47-62.

Williams-Linera G. 1997. Phenology of deciduous and broadleaved-evergreen tree species in a Mexican tropical lower montane forest. *Global Ecology and Biogeography Letters* **6**: 115-127.

Winkler M. & Hietz P. 2001. Population structure of three epiphytic orchids (*Lycaste aromatica*, *Jacquiniella leucomelana*, and *J. teretifolia*) in a Mexican humid montane forest. *Selbyana* **22**: 27-33.

Zotz G. & Hietz P. 2001. The physiological ecology of vascular epiphytes: Current knowledge, open questions. *Journal of Experimental Botany* **52**: 2067-2078.

Zotz G. & Vollrath B. 2002. Substrate preferences of epiphytic bromeliads: An experimental approach. *Acta Oecologica* **23**: 99-102.

Zotz G. & Ziegler H. 1999. Size-related differences in carbon isotope discrimination in the epiphytic orchid, *Dimerandra emarginata*. *Naturwissenschaften* **86**: 39-40.

Zotz G. 1997. Substrate use of three epiphytic bromeliads. *Ecography* **20**: 264-270.

Zotz G. 1998. Demography of the epiphytic orchid, *Dimerandra emarginata*. *Journal of Tropical Ecology* **14**: 725-741.

2 Herbivory in epiphytic bromeliads, orchids and ferns in a Mexican montane forest

Manuela Winkler, Karl Hülber, Klaus Mehltreter, José García Franco and Peter Hietz

published in Journal of Tropical Ecology (2005) 21: 147-154

Copyright © 2005, Cambridge Journals (reprinted with permission)

Abstract

Herbivory is important in tropical woody plants, but the few data available suggest that rates of herbivory are mostly low in epiphytes. We quantified herbivory at the community level in five bromeliad, three orchid and five fern species of a Mexican humid montane forest. Leaf area loss was < 1.5 % in bromeliads and orchids, but much higher (7 - 20 %) in ferns. Damage was positively correlated with leaf nitrogen content but not with leaf life span. In contrast to low leaf damage, many bromeliads were infested by curculionid larvae feeding on the meristematic tissue at the ramet base, and we estimate that this accounts for 18 and 31 % of ramet and shoot death of large individuals of *Tillandsia punctulata* and *T. deppeana*, respectively. Herbivory in flowers, capsules or inflorescence stalks reduced fecundity by *ca.* 14 - 18 % in three of the five bromeliads and by 90 % in the orchid *Lycaste aromatica*, but had little effect on the other species. These data show that even if the leaf area consumed is indeed low in epiphytic orchids and bromeliads, the less conspicuous damage done to reproductive organs and meristematic tissue can have a strong effect on fecundity and survival.

Key words: Bromeliaceae, ferns, herbivory, humid montane forest, Mexico, Orchidaceae, Polypodiaceae

2.1 Introduction

Epiphytes play important roles in ecosystem processes of tropical rain forests and contribute substantially to their biodiversity by providing habitat and other resources for the fauna living in the forest canopy (Benzing 1990, Frank 1983). Whereas the interactions between epiphytes and pollinators have often been studied (Ackerman 1986, Bartareau 1995, Zimmerman *et al.* 1989), reports on the effect of herbivores on epiphytes are very scarce (e.g. Lowman *et al.* 1996, Schmidt & Zotz 2002). This contrasts with a large body of detailed studies on herbivory in tropical shrubs and trees (Coley 1983, Coley & Barone 1996, Landsberg & Ohmart 1989, Lowman 1984, 1985; Newbery & de Foresta 1985). Some studies sampled in the tree canopy, but very few included life forms other than trees (Sterck *et al.* 1992).

Ecological theory predicts that species with long-lived leaves are more common in resource-poor habitats where the replacement of leaves is expensive (Coley & Aide 1991, Coley *et al.* 1985, reviewed in Landsberg & Ohmart 1989). Because leaves with longer life spans face a greater risk of being found and eaten, they are generally better defended than shorter-lived leaves (Coley & Aide 1991, Coley & Barone 1996, Coley *et al.* 1985). Since the epiphytic habitats are often resource poor (Benzing 1990, Zotz & Hietz 2001), herbivory levels in epiphytes can be expected to be relatively low (Coley *et al.* 1985). Moreover, the low nitrogen content (Zotz & Hietz 2001) and the fact that the leaves of many species, particularly bromeliads, are tough, presumably with a high fibre content, suggests low nutritional value and palatability. A detailed study of the epiphytic bromeliad *Vriesea sanguinolenta* found a mean annual leaf area consumed per plant between 4.1 and 8.4 % and between and 26 to 61 % of individuals affected by herbivory, challenging the notion of low epiphyte herbivory (Schmidt & Zotz 2000).

However, herbivores also feed on organs other than leaves. Several weevils (Curculionidae) deposit eggs into slits cut into leaf bases of bromeliads, their larvae develop in the meristematic tissue at the base of the ramets (Frank & Thomas 1994). One species (*Metamasius callizona*) was introduced to Florida from eastern Mexico and, apparently without natural enemies, seriously endangers native bromeliad populations (Frank & Thomas 1991, 1994). The effect of these species on bromeliads in their native habitat has not received the same attention and to our knowledge has never been quantified. Pierce & Gottsberger (2001) reported a number of floriphagous weevil species on bromeliad flowers, also without quantifying the effect on the plants.

Questioning generalizations made from observations of single species or organs, we assessed herbivory in several species of bromeliad, orchid and fern, and distinguished the damage of different plant parts. As this study was part of a project on epiphyte population dynamics, we attempted to evaluate any effect to the populations by sampling individuals of different size classes, from different canopy positions and at different times of the year.

2.2 Methods

Study site

This study was conducted in a small forest reserve adjacent to the Instituto de Ecología, 2.5 km south of Xalapa, in central Veracruz, Mexico (19°31'N, 96°57'W), at 1350 m elevation. Average temperature is 19° C, and annual precipitation is 1500 mm, most of which falls in the June-October wet season. According to the Holdridge life-zone system (Holdridge 1967), the forest is at the transition between premontane and lower montane moist forest. In Mexico, it is commonly classified as 'bosque mesófilo de montaña' (mesophilous montane forest) according to Rzedowski (1986). Descriptions of the forest structure are given by Williams-Linera (1997) and of the epiphyte community by Hietz & Hietz-Seifert (1995).

Study species

Herbivory was recorded in five fern (*Pleopeltis crassinervata* (Fée) T. Moore, *Polypodium furfuraceum* Schltdl. & Cham., *P. plebeium* Schltdl. & Cham., *P. polypodioides* (L.) Watt and *P. rhodopleuron* Kunze, all Polypodiacae), five bromeliad (*Catopsis sessiliflora* (Ruiz & Pavón) Mez, *Tillandsia deppeana* Steudel, *T. juncea* (Ruiz & Pav.) Poir., *T. multicaulis* Steud. and *T. punctulata* Schltdl. & Cham.) and three orchid species (*Jacquiniella leucomelana* (Reichenbach f.) Schlechter, *J. teretifolia* (Sw.) Britton & P. Wilson and *Lycaste aromatica* (Graham ex Hook.) Lindley), representing the most important taxonomic groups of neotropical epiphytes and the most common epiphytic species at our study site.

Folivory

For the bromeliads and orchids, we selected 250 sections of branches on 16 trees, representing the range of canopy strata colonised by the epiphytes studied. All plants on these sections were tagged, the length of the longest leaf (bromeliads), pseudobulb height (*L. aromatica*) or ramet length (*Jacquiniella* spp.) was measured and folivory was estimated visually for each individual and assigned to eight classes (0, 1-2, 3-5, 6-10, 11-25, 26-50, 51-75 and 76-100 % leaf area loss). For all calculations, the class median was used.

Identifying missing leaf area as consumed is not always possible and some loss may result from mechanical damage (e.g. by falling branches), wilting or microbial infections (Coley 1983). At the first census in February/March 2002 the total missing leaf area was recorded. In the following two censuses (rainy season in August/September and dry season 2003), a more conservative approach was taken and only damage with feeding signs was considered.

Folivory in the fern species was estimated only at the beginning of the dry season in 2003. One hundred and ten individuals (*i.e.* fronds growing on one rhizome) on 66 branch sections were assessed. The proportion of each pinna (or frond in the case of the undissected *P. crassinervata*) consumed was estimated in situ in the same classes as for orchids and bromeliads except that damage classes 2, 5 and 10 % were pooled. The class median (e.g., the median of the 10-25 % class is 17.5) of each pinna was used to calculate the mean damage per frond (Mehltreter & Tolome 2003).

Because leaves of some plants live for several years and at least bromeliad ramets have determinate growth where all leaves die at once, it did not appear appropriate to estimate leaf life span by observing a larger number of leaves from unfolding to death. Instead, leaves of fern individuals and bromeliad and orchid ramets were counted and marked with a water-based ink, and newly produced leaves were counted after 10 mo in *J. leucomelana* (59 ramets, 27 individuals), *J. teretifolia* (84, 29) and *C. sessiliflora* (24, 21), after 11 mo in *T. multicaulis* (26, 15), and after 12 mo in *P. crassinervata* (22 individuals), *P. furfuraceum* (19), *P. plebeium* (30), and *P. polypodioides* (14). Ramets that flowered in the observation period where excluded. Leaf life span was calculated similar to Tanner (1983) as:

total number of living leaves present initially / number of new leaves x time span (mo).

Foliar nitrogen concentration was measured in continuous-flow mode in an elemental analyzer (EA 1110, CE Instruments, Italy) (Rammler 2004 and unpublished data).

Herbivory in reproductive organs

For all fertile bromeliad and orchid individuals on the branch sections, the percentage of damaged flowers and fruits was estimated with the same classes and time intervals as for folivory. Damage to an inflorescence stalk normally destroys the whole distal inflorescence and all flowers and fruits distal to the damaged point were regarded as lost.

Additionally, in 415 randomly selected fertile bromeliad individuals beyond the tagged branch sections (91/57 *C. sessiliflora* female/male, 38 *T. deppeana*, 41 *T. juncea*, 91 *T. multicaulis* and 97 *T. punctulata*) flowers and fruits were counted and herbivory was recorded during the year 2002. In these plants only the flowers distal to a damaged region on inflorescence stalks were considered lost. If the inflorescence was too heavily damaged to count the affected flowers, the average number of fruits per inflorescence of the respective species was calculated as lost.

Herbivory in bromeliad ramets

Bromeliads that fell either with their branch or bark support, or from their support were collected from five 20 x 20-m plots around the sampled trees every second month from October 2002 to April 2003. All fallen ramets above a size threshold (*C. sessiliflora* lengths of longest leaf $\geq$ 15 cm, *T. deppeana* $\geq$ 35 cm, *T. juncea* $\geq$ 40 cm, *T. multicaulis* and *T. punctulata* $\geq$ 20 cm) were opened to check for herbivores (mining larvae) or signs of herbivory (galleries, frass). Smaller ramets are much less likely to be hosts as weevils require a minimal stem diameter to mature (Frank 1999, Frank & Sidoti 2002). We assumed that if a plant falls with its support this will be independent of herbivore damage and that infestation rates in these plants are therefore representative of the whole population, but that at least some plants will fall from their support because they have been weakened or killed by herbivores. Supposing that the different infestation rates observed in plants fallen with and without their support is a result of infestation, we calculated the proportion of plants falling because of infestation as

$$(N_{Hs}-N_{HS} / N_{hS} \times N_{hs}) / N$$

where N is the total number of ramets, the index H indicates herbivory in the ramet base, h no herbivory, S fallen with support, and s fallen without support.

Statistical analysis

Differences between groups were tested with Kruskal-Wallis-test (comparison of mean folivory in fern, bromeliad and orchid species), Wilcoxon matched-pairs test (comparison of folivory in different seasons) or χ^2-test (to test for differences in survival between plants with and without herbivory, comparison of infestation rate in ramets fallen with or without their support). Correlations between leaf life span, leaf nitrogen content and mean leaf area missing and between size and missing leaf area were calculated with Spearman rank correlations. Statistical tests were calculated using SPSS 9.0 (SPSS Inc., 1989-1999).

2.3 Results

Folivory

At the beginning of the dry season of 2003, average leaf area loss did not exceed 1.3 % in the orchids and bromeliads, but four ferns exhibited mean damage of ca. 10 %, and in *P. plebeium* 20.3 % of the leaf area was missing (Figure 2.1a). Differences between the species means of these three taxonomic groups were statistically significant (Kruskal-Wallis test: $\chi^2 = 9$, df = 2, P = 0.009). Between 60.7 % and 95 % of the sampled fern fronds and all individuals were affected by folivory, compared to a maximum of 31.7 % of orchid (*L. aromatica*) and 15 % of bromeliad (*T. punctulata*) individuals (Figure 2.1b). There were no significant differences in folivory between rainy season 2002 and dry season 2003, except in *T. punctulata* (Wilcoxon matched pairs test: Z = -2.3, P = 0.021), *J. leucomelana* (Z = -4.2, P < 0.001) and *J. teretifolia* (Z = -2.2, P = 0.028). The values were below 1.5 % in these species in both seasons. Missing leaf area in February/March 2002 was higher than in the same season in 2003, when we attempted to exclude damage due to wilting or mechanical damage in bromeliads and orchids (Table 2.1). Bromeliads and orchids affected by folivory in 2002 had equal or even lower mortality than plants without damage (Table 2.2).

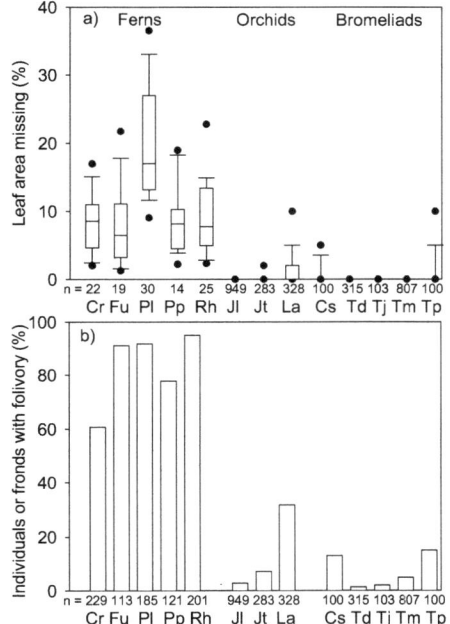

Figure 2.1: Folivory at the beginning of the dry season (February/March) 2003. (a) Missing leaf area per individual (median, quartiles, 10th and 90th percentiles, the dots are the 5th and 95th percentiles). (b) Proportion of individuals (bromeliads and orchids) or fronds (ferns) with folivory. Ferns: Cr = *Pleopeltis crassinervata*, Fu = *Polypodium furfuraceum*, Pl = *P. plebeium*, Pp = *P. polypodioides*, Rh = *P. rhodopleuron*, .Orchids: Jl = *Jacquiniella leucomelana*, Jt = *J. teretifolia*, La = *Lycaste aromatica*. Bromeliads: Cs = *Catopsis essiliflora*, Tj = *Tillandsia juncea*, Tp = *T. punctulata*, Td = *T. deppeana*, Tm = *T. multicaulis*.

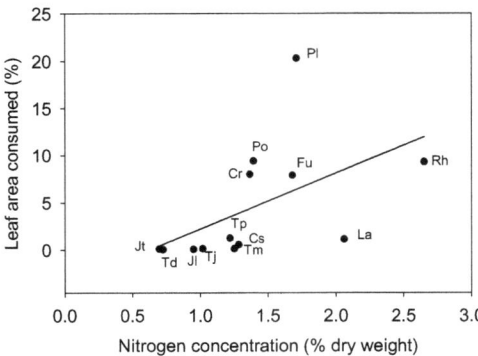

Figure 2.2: Leaf nitrogen content and mean leaf area consumed with linear regression line. Species codes as in Figure 2.1.

Table 2.1: Percentage of mean leaf area lost per individual assigned to folivory, which is derived from a comparison of the missing leaf area between the dry season of 2002 (total missing leaf area recorded) and the dry season of 2003 (excluding obvious mechanical and wilting damage), Z- and P-values of Wilcoxon matched-pairs test for differences between dry season 2002 and 2003, N is the number of individuals.

Species	Folivory / total leaf area loss (%)	Z	P	N
Bromeliaceae				
Catopsis sessiliflora	55	-1.6	0.108	76
Tillandsia deppeana	73	-1.2	0.219	12
T. juncea	20	-3.0	0.003	101
T. multicaulis	41	-4.5	0.000	143
T. punctulata	65	-1.0	0.328	99
Orchidaceae				
Jacquiniella leucomelana	52	-2.9	0.004	666
J. teretifolia	60	-2.0	0.045	280
Lycaste aromatica	40	-5.7	0.000	317
Total (including *T.* spp juveniles)	45	-8.7	0.000	2134

Table 2.2: Percentage of individuals with and without folivory in the rainy season 2002 surviving to the dry season 2003. Groups were tested for differences with χ^2-test. All *Tillandsia* spp. are considered as one group because very small juveniles could not be determined to species level.

Species	Percentage of individuals surviving		χ^2-test		
	with folivory	without folivory	χ^2	P	N
Bromeliaceae					
Catopsis sessiliflora	78.6 %	75.0 %	0.08	0.772	110
Tillandsia spp.	92.5 %	64.0 %	18.3	<0.001	2142
Orchidaceae					
Jacquiniella leucomelana	94.8 %	86.3 %	4.48	0.034	845
Jacquiniella teretifolia	76.3 %	80.3 %	0.34	0.561	348
Lycaste aromatica	98.3 %	91.6 %	6.06	0.014	342

Nitrogen content in epiphyte leaves ranged from 0.7 to 2.7 % dry weight (Table 2.3). Mean leaf area lost was positively correlated with leaf nitrogen content (Spearman rho: 0.81, P = 0.001, Figure 2.2), mainly because of the high damage and high nitrogen concentration in ferns. Mean leaf life span ranged from 19 to 32 mo, except for the drought-deciduous leaves of *L. aromatica* and *P. rhodopleuron*, which last about 9 mo (Table 2.3). There was no significant correlation between leaf life span and leaf damage. The proportion of leaf area consumed increased with plant size in bromeliads (*Tillandsia* spp.: r = 0.32, P < 0.001, *C. sessiliflora*: r = 0.38, P < 0.001), and orchids (*J. leucomelana*: r = 0.12, P < 0.001, *J. teretifolia*: r = 0.16, P = 0.008, *L. aromatica*: r = 0.24, P < 0.001), but not in ferns (*P. furfuraceum*: r = -0.43, *P. plebeium*: r = -0.05, *P. polypodioides*: r = -0.12, *P. rhodopleuron*: r = -0.06; all P > 0.05).

Table 2.3: Foliar nitrogen concentration and leaf life span (mean ± SD [N]) of epiphytes in the forest of the Instituto de Ecología, Jalapa, Mexico.

Species	Nitrogen concentration (% dry weight)	Leaf life span (mo)
Polypodiaceae		
Pleopeltis crassinervata	1.4 ± 0.35 (78)	25 ± 21.6 (21)
Polypodium furfuraceum	1.7 ± 0.25 (3)	21 ± 11.2 (19)
P. plebeium	1.7 ± 0.41 (15)	32 ± 22.0 (30)
P. polypodioides	1.4 ± 0.42 (2)	22 ± 20.3 (14)
P. rhodopleuron	2.7 (1)	9
Bromeliaceae		
Catopsis sessiliflora	1.3 ± 0.5 (17)	23 ± 9.6 (21)
Tillandsia deppeana	0.7 ± 0.24 (9)	
T. juncea	1.0 ± 0.28 (30)	
T. multicaulis	1.3 ± 0.47 (10)	31 ± 11.7 (15)
T. punctulata	1.2 ± 0.33 (28)	
Orchidaceae		
Jacquiniella leucomelana	1.0 (1)	19 ± 12.1 (27)
J. teretifolia	0.7 ± 0.06 (3)	23 ± 7.7 (29)
Lycaste aromatica	2.1 ± 0.54 (5)	9

Herbivory in reproductive organs

In *C. sessiliflora*, *T. deppeana* and *T. juncea*, herbivory in reproductive plant parts differed considerably between rainy and dry season in the field whereas *T. multicaulis* and *T. punctulata* showed no seasonal variation (Figure 2.3a). Herbivory in inflorescence stalks alone (without herbivory in single flowers) reduced fecundity by ca. 14-18 % in *T. juncea*, *T. deppeana* and *T. punctulata* (Figure 2.3b). Herbivory in flowers or fruits was never observed in *Jacquiniella* spp. By contrast, 90 % of the fruits of *L. aromatica* were infested and failed to produce seeds by the dry season of 2003.

Herbivory in bromeliad ramets

The proportion of infested ramets was highest in *T. deppeana*, followed by *T. punctulata*, *T. multicaulis*, *C. sessiliflora* and *T. juncea* (Figure 2.4). Infestation was much higher in plants that fell without their support than in those that fell with it in all species. We estimate that herbivore infestation was responsible for 1 % of ramets falling to the ground and subsequently dying in *C. sessiliflora* and *T. juncea*, 3 % in *T. multicaulis*, 18 % in *T. punctulata* and 31 % in *T. deppeana*. In the meristematic tissue of several ramets of *T. deppeana* coleoptera larvae were found, two of which metamorphed and were identified as *Metamasius sellatus* (Curculionidae, Coleoptera).

2.4 Discussion

Folivory

Assessing whether a hole in a leaf or a missing leaf tip was caused by herbivores or not can be difficult. We therefore estimated the total leaf area missing in the first observation and applied a more conservative approach for the following seasons by excluding damage obviously not caused by herbivores. Herbivory rates can differ between years, but as we recorded no significant differences between rainy and dry season, which are likely to differ in leaf growth and

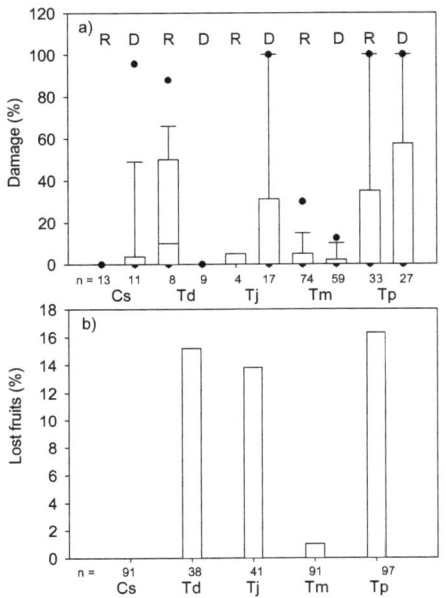

Figure 2.3: Herbivory in bromeliad inflorescences. (a) Estimated proportion of damaged flowers/fruits per individual (median, quartiles, 10th and 90th percentiles, the dots are the 5th and 95th percentiles) of all individuals on marked branch sections. R: rainy season 2002, D: dry season 2003. (b) Proportion of potential fruit production lost due to herbivory in inflorescence stalks but excluding damage to individual flowers. Data derived from flower/fruit counts in randomly selected individuals. Species codes as in Figure 2.1.

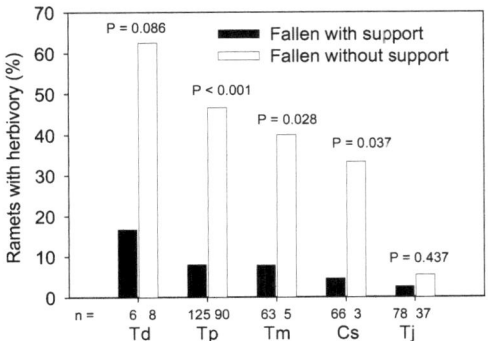

Figure 2.4: Percentage of infested bromeliad ramets, compared between ramets fallen with or without their support (branch or bark) with χ^2-test (*C. sessiliflora* $\chi^2 = 4.9$, *T. deppeana* $\chi^2 = 3.6$, *T. juncea* $\chi^2 = 8.5$, *T. multicaulis* $\chi^2 = 27.1$, *T. punctulata* $\chi^2 = 67.9$). Species codes as in Figure 2.1.

herbivore activity, we assume folivory in the dry seasons of 2002 and 2003 to be similar. In both cases, missing leaf area was low in the orchids and bromeliads. Less than half of the total damage could be attributed to folivory (Table 2.1), which is consistent with Barone (2000) who found that in the tropical tree *Alseis blackiana* pathogens and physical damage accounted for ca. 35 % of the missing leaf area. Since mortality in orchids and bromeliads with folivory was not higher than in unaffected plants (Table 2.2), the low rates of leaf loss in these species do not appear to affect survival. The higher mortality of some species in plants without folivory can be explained by the fact that large plants, with generally lower mortality (Hietz 1997), were more likely to show at least minimal damage. In general, the low herbivory observed in orchids and bromeliads accords with the notion that epiphytes experience relatively little herbivory (e.g. Benzing 1990, Zotz 1998). However, Schmidt & Zotz (2000) found mean annual rates of folivory per plant of 4.1 – 8.4 % in the epiphytic bromeliad *Vriesea sanguinolenta* causing a mean herbivore-induced mortality rate of 2.2 % per year, and Schmidt & Zotz (2002) report that 8.1 % of total mortality in *Vriesea sanguinolenta* but none in *Aspasia principissa* (Orchidaceae) was induced by herbivory.

In other plant groups, folivory is often much higher. Filip *et al.* (1995) report an overall mean of 7.6 % leaf area loss in 16 tropical tree species. In ten tree species of different plant families and of different successional status in Papua New Guinea folivory averaged 3.8 – 19.7 % (Basset & Höft 1994). 2.9–7.5 % of the leaf area of young leaves of a neotropical liana were damaged by herbivores (Aide & Zimmerman 1990). Mean annual rates of leaf damage in temperate forests are 7.1 %, in tropical wet forests 11.1 % for shade-tolerant tree species and 48 % for gap specialists, and in tropical dry forests 14.2 % (reviewed in Coley & Barone 1996). In the subtropical humid montane forest we worked in, leaf area loss to herbivores was ca. 4 % in a temperate deciduous tree and 3-8 % in two tropical evergreen tree species (Williams-Linera & Baltazar 2001)

Missing leaf area was more than 20 % in *Polypodium plebeium* and 8-10 % in four other fern species. Even if this over-estimates folivory, the leaf area loss caused by herbivores (unidentified caterpillars, apparently of one species, were observed feeding on several fern species) is certainly much greater than in the bromeliads and orchids observed. These data are consistent with substantial herbivory in terrestrial tropical ferns (Balick *et al.* 1978, Hendrix & Marquis 1983, Koptur *et al.* 1998, Mehltreter & Tolome 2003, Shuter & Westoby 1992)

Longer-lived leaves tend to have lower palatability and are better defended against herbivores (Coley 1983, Coley et al. 1985, Southwood et al. 1986). In saplings of 46 rain-forest tree species leaf toughness had the highest correlation with folivory, followed by fibre content and nutritive value (Coley 1983). Ferns with similar leaf longevity as bromeliads and orchids experienced a remarkably higher rate of damage. Compared to fern fronds, bromeliad leaves are hard and fibrous, and nitrogen concentration in ferns was higher than in bromeliads and evergreen orchids. Mean leaf area lost was positively correlated with leaf nitrogen concentration. Thus, unless protected by toxic compounds or ants, the nutritional value and palatability should make fern fronds more attractive to herbivores than bromeliad or orchid leaves.

Herbivory in reproductive organs

Herbivory considerably reduced fecundity in *Tillandsia deppeana*, *T. punctulata* and *T. juncea*, where a large proportion of the loss of flowers was caused by damage to inflorescence stalks, which are more exposed than those of *T. multicaulis* and thicker than those of *Catopsis sessiliflora*. However, relative to the high number of seeds failing to reach safe sites for germination (Benzing 1978, Castro-Hernández et al. 1999), the observed reduction in fecundity may not be limiting population growth.

In the three orchids the proportion of damaged fruits differed dramatically between 90 % in *Lycaste aromatica* and no damage in *Jacquiniella* spp. This may be the cause for the low recruitment in *Lycaste* compared to the two *Jacquiniella* species (Winkler & Hietz 2001).

Herbivory in bromeliad ramets

Insects feeding on meristematic tissue of bromeliad ramets can kill them (Frank & Thomas 1991, 1994). We found bromeliads fallen without their supporting branch more frequently infested, suggesting that herbivore attack caused the bromeliad ramets to fall and thus contributes significantly to mortality. Since, with the possible exception of fruiting *T. deppeana*, dying bromeliads rarely disintegrate on the branch, but fall to the ground as entire ramets, those found on the ground constitute the bulk of mortality. *Tillandsia deppeana*, the species most affected by herbivory at the base, is largely monocarpic and loss of the shoot equals death of the whole plant.

The other species have several ramets, not all of which will be infested. Herbivory at the ramet base contributes 31 % to total mortality in *T. deppeana* and 18 % to ramet mortality in *T. punctulata*. The latter species has an average of six ramets per individual, but the loss of any ramet may affect future reproduction or growth (Juenger & Bergelson 1997, Marquis 1996, Sagers & Coley 1995).

Conclusions

The rates of folivory among the observed bromeliads and orchids accord with the suggested low herbivory in vascular epiphytes, but leaf area lost in ferns was substantial. In contrast, damage done to the meristematic tissue, although not readily visible, is an important cause of mortality for large bromeliads. Damage done to reproductive organs differs strongly among species, substantially reducing fecundity in the bromeliads and even more so for *Lycaste aromatica*.

2.5 Acknowledgements

We thank Leticia Cruz Paredes and Angélica Jimenez Aguilar for help in the field, Howard Frank for identifying curculionids and Wolfgang Wanek for measuring N content. We are grateful to the botanical garden Francisco Clavijero and the Instituto de Ecología in Xalapa for general support. Gerhard Zotz and three anonymous reviewers provided useful comments on earlier drafts. This research was funded by the Austrian Science Fund (FWF grant number P14775).

2.6 Literature

Ackerman J. D. 1986. Coping with the epiphytic existence: pollination strategies. *Selbyana* **9**: 52-60.

Aide T. M. & Zimmerman J. K. 1990: Patterns of insect herbivory, growth, and survivorship in juveniles of a neotropical liana. *Ecology* **71**: 1412-1421.

Balick M. J., Furth D. G. & Cooper-Driver G. 1978. Biochemical and evolutionary aspects of arthropod predation on ferns. *Oecologia* 35: 55-89.

Barone J. A. 2000. Comparison of herbivores and herbivory in the canopy and understory for two tropical tree species. *Biotropica* **32**: 307-317.

Bartareau T. 1995. Pollination limitation, costs of capsule production and the capsule-to-flower ratio in *Dendrobium monophyllum* F. Muell., (Orchidaceae). *Australian Journal of Ecology* **20**: 257-265.

Basset Y. & Höft R. 1994. Can apparent leaf damage in tropical trees be predicted by herbivore load on host-related variables? A case study in Papua New Guinea. *Selbyana* **15**: 3-13.

Benzing D. H. 1978. Germination and early establishment of *Tillandsia circinnata* Schlecht. (Bromeliaceae) on some of its hosts and other supports in southern Florida. *Selbyana* **5**: 95-106.

Benzing D. H. 1990. *Vascular epiphytes – general biology and related biota*. Cambridge: Cambridge University Press. 354 pp.

Castro-Hernández J. C., Wolf J. H., Garcia-Franco J. G. & Gonzalez-Espinosa M. 1999. The influence of humidity, nutrients and light on the establishment of the epiphytic bromeliad *Tillandsia guatemalensis* in the highlands of Chiapas, Mexico. *Revista de Biología Tropical* **47**: 763-773.

Coley P. D. 1983. Herbivory and defense characteristics of tree species in a lowland tropical forest. *Ecological Monographs* **53**: 209-233.

Coley P. D. & Aide T. M. 1991. Comparison of herbivory and plant defenses in temperate and tropical broad-leaved forests. Pp 25-49 in Price P. W., Lewinsohn T. M., Fernandes G. W. & Benson W. W. (eds.). *Plant-animal interactions: evolutionary ecology in tropical and temperate regions*. New York: John Wiley & Sons.

Coley P. D. & Barone J. A. 1996. Herbivory and plant defenses in tropical forests. *Annual Review of Ecology and Systematics* **27**: 305-335.

Coley P. D., Bryant J. P. & Chapin F. S. 1985. Resource availability and plant antiherbivore defenses. *Science* **230**: 895-899.

Filip V., Dirzo R., Maass J. M. & Sarukhán J. 1995. Within- and among-year variation in the levels of herbivory on the foliage of trees from a Mexican tropical deciduous forest. *Biotropica* **27**: 78-86.

Frank J. H. 1983. Bromeliad phytotelmata and their biota, especially mosquitoes. Pp 101-128 in Frank J. H. & Lounibos L. P. (eds.). *Phytotelmata: terrestrial plants as hosts for aquatic insect communities*. Medford: Plexus Publishing Inc.

Frank J. H. 1999. Bromeliad-eating weevils. *Selbyana* **20**: 40-48.

Frank J. H. & Sidoti B. J. 2002. The effect of size of host plant (*Tillandsia utriculata*. Bromeliaceae) on development of *Metamasius callizona* (Dryophthoridae). *Selbyana* **23**: 220-223.

Frank J. H. & Thomas M. C. 1991. *Metamasius callizona* kills bromeliads in southeastern Florida. *Journal of the Bromeliad Society* **41**: 107–108.

Frank J. H. & Thomas M. C. 1994. *Metamasius callizona* (Chevrolat) (Coleoptera. Curculionidae), an immigrant pest, destroys bromeliads in Florida. *The Canadian Entomologist* **126**: 673-682.

Hendrix S. D. & Marquis R. J. 1983. Herbivore damage to three tropical ferns. *Biotropica* **15**: 108-111.

Hietz P. 1997. Population dynamics of epiphytes in a Mexican humid montane forest. *Journal of Ecology* **85**: 767-775.

Hietz P. & Hietz-Seifert U. 1995. Intra- and interspecific relations within an epiphyte community in a Mexican humid montane forest. *Selbyana* **16**: 135-140.

Holdridge L. R. 1967. *Life zone ecology*. San José, Costa Rica: Tropical Science Center. 206 pp.

Juenger T. & Bergelson J. 1997. Pollen and resource limitation of compensation to herbivory in scarlet gilia, *Ipomopsis aggregata*. *Ecology* **78**: 1684-1695.

Koptur S., Rico-Gray V. & Palacios-Rios M. 1998. Ant protection of the nectaried fern *Polypodium plebeium* in central Mexico. *American Journal of Botany* **85**: 736-739.

Landsberg J. & Ohmart C. 1989. Levels of insect defoliation in forests. Pattern and concepts. *Trends in Ecology and Evolution* **4**: 96-100.

Lowman M. D. 1984. An assessment of techniques for measuring herbivory. Is rain forest defoliation more intense than we thought? *Biotropica* **16**: 264-268.

Lowman M. D. 1985. Insect herbivory in Australian rain forests – is it higher than in the Neotropics? *Proceedings of the Ecological Society of Australia* **14**: 109-119.

Lowman M. D., Wittman P. K. & Murray D. 1996. Herbivory in a bromeliad of the Peruvian rain forest canopy. *Journal of the Bromeliad Society* **46**: 52-55.

Marquis R. J. 1996. Plant architecture, sectoriality and plant tolerance to herbivores. *Vegetatio* **127**: 85-97.

Mehltreter K. & Tolome J. 2003. Herbivory on three tropical fern species of a Mexican cloud forest. Pp. 375 - 381 in Chandra S. & Srivastava M. (eds.). *Pteridology in the new millennium*. The Netherlands: Kluwer Academic Publishers.

Newbery D. M. & De Foresta H. 1985. Herbivory and defense in pioneer, gap and understory trees of tropical rain forest in French Guiana. *Biotropica* **17**: 238-244.

Pierce S. & Gottsberger R. A. 2001. Bromeliad flowers, an attractive meal for weevils at Cerro Jefe, Panama. *Journal of the Bromeliad Society* **51**: 172-176.

Rammler H. 2004. *Ökophysiologie von Epiphyten und Bäumen in einem Bergregenwald in Mexiko und einem Tieflandregenwald in Costa Rica*. Diploma Thesis, Universität für Bodenkultur, Vienna. 106 pp.

Rzedowski J. 1986. *Vegetación de México*. (Third edition). México: Editorial Limusa. 432 pp.

Sagers C. L. & Coley P. D. 1995. Benefits and costs of defense in a neotropical shrub. *Ecology* **76**: 1835-1843.

Schmidt G. & Zotz G. 2000. Herbivory in the epiphyte, *Vriesea sanguinolenta* Cogn. & Marchal (Bromeliaceae). *Journal of Tropical Ecology* **16**: 829-839.

Schmidt G. & Zotz G. 2002. Inherently slow growth in two Caribbean epiphytic species: a demographic approach. *Journal of Vegetation Science* **13**: 527-534.

Shuter E. & Westoby A. 1992. Herbivorous arthropods on bracken *Pteridium aquilinum* (L.) Kuhn in Australia compared with elsewhere. *Australian Journal of Ecology* **17**: 329-339.

Southwood T. R., Brown V. K. & Reader P. M. 1986. Leaf palatability, life expectancy and herbivore damage. *Oecologia* **70**: 544-548.

Sterck F., van der Meer P. & Bongers F. 1992. Herbivory in two rain forest canopies in French Guiana. *Biotropica* **24**: 97-99.

Tanner E. V. J. 1983. Leaf demography and growth of the tree-fern *Cyathea pubescens* Mett. ex Kuhn in Jamaica. *Botanical Journal of the Linnean Society* **87**: 213-227.

Williams-Linera G. 1997. Phenology of deciduous and broadleaved-evergreen tree species in a Mexican tropical lower montane forest. *Global Ecology and Biogeography Letters* **6**: 115-127.

Williams-Linera G. & Baltazar A. 2001. Herbivory on young and mature leaves of one temperate deciduous and two tropical evergreen trees in the understory and canopy of a Mexican cloud forest. *Selbyana* **22**: 213-218.

Winkler M. & Hietz P. 2001. Population structure of three epiphytic orchids (*Lycaste aromatica, Jacquiniella leucomelana,* and *J. teretifolia*) in a Mexican humid montane forest. *Selbyana* **22**: 27-33.

Zimmerman J. K., Roubik D. W. & Ackerman J. D. 1989. Asynchronous phenologies of a neotropical orchid and its euglossine bee pollinator. *Ecology* **70**: 1192-1195.

Zotz G. 1998. Demography of the epiphytic orchid, *Dimerandra emarginata*. *Journal of Tropical Ecology* **14**: 725-741.

Zotz G. & Hietz P. 2001. The physiological ecology of vascular epiphytes: current knowledge, open questions. *Journal of Experimental Botany* **52**: 2067-2078.

3 Effect of canopy position on germination and seedling survival of epiphytic bromeliads in a Mexican humid montane forest

Manuela Winkler, Karl Hülber and Peter Hietz

published in Annals of Botany (2005) 95: 1039-1047

© The Author 2005. Published by Oxford University Press on behalf of the Annals of Botany Company (reprinted with permission)

Abstract

Seeds of epiphytes have to land on branches with suitable substrates and microclimates to germinate and for the resulting seedlings to survive. It is important to understand the fate of seeds and seedlings to model populations, but it is often neglected when only established plants are included.

We exposed the seeds of five bromeliad species in different canopy positions and recorded germination and early seedling survival in a Mexican montane forest. Additionally, the survival of naturally dispersed seedlings was monitored over 2.5 years. We used survival analysis, a procedure rarely used in plant ecology, to study the influence of branch characteristics and light on germination and seedling survival in natural and experimental populations.

Experimental germination percentages ranged from 7.2 % in *Tillandsia deppeana* to 33.7 % in *T. juncea*, but the seeds of *T. multicaulis* largely failed to germinate. Twenty months after exposure between 3.5 and 9.4 % of the seedlings were still alive. There was no evidence that canopy position affected the probabilities of germination, but time to germination was shorter in less exposed canopy positions indicating that higher humidity accelerates germination. More experimental seedlings survived when canopy openness was high, whereas survival in census seedlings was influenced by moss cover. While mortality decreased steadily with age in the atmospheric *Tillandsia* juveniles, in the more mesomorphic *Catopsis sessiliflora* mortality increased dramatically in the dry season.

Seedling mortality, rather than the failure to germinate, accounts for the differential distribution of epiphytes within the canopy studied. With few safe sites to germinate and high seedling mortality, changes of local climate may affect epiphyte populations primarily through their seedling stage.

Key words: Bromeliaceae, *Catopsis*, epiphytes, germination, habitat heterogeneity, Mexico, seedling establishment, survival analysis, *Tillandsia*

3.1 Introduction

Tree crowns impose a powerful set of constraints to epiphytes, offering a fragmented and stressful habitat with substantial rates of patch turnover (Benzing 1990). Epiphytes experience tree crowns as a mosaic of suitable and unsuitable habitat, with the suitable type occurring in relatively small and discrete patches (Hanski & Gilpin 1997). Seed size in epiphytes rarely exceeds 2 mm regardless of the dispersal mechanism, and 84 % of epiphyte species are adapted to wind dispersal either by dust-like diaspores or by winged or plumed seeds (Madison 1977). Small seed size is advantageous for epiphytes as the potential habitat can be showered with numerous seeds, increasing the chance that an adequate number will fall onto safe sites. Furthermore, small seeds are more likely to slip into cracks in bark and their large surface to volume ratio favours the rapid uptake of water (Madison 1977). On the other hand, small seeds provide few resources for seedlings to succeed on bark substrates (Benzing 1990). Some epiphytes have developed special devices for transport and attachment to bark (van der Pijl 1972). For instance, members of the bromeliad subfamily Tillandsioideae disperse via small seeds equipped with coma hairs extending from one or both ends of the integument (Benzing 2000). These constitute more than half of the aggregate seed mass in epiphytic species, providing buoyancy and holdfast to substrates (Benzing 2000). Seedling mortality can be expected to be high because of small seed size and high proportional investment in the dispersal apparatus (Howe & Smallwood 1982, Jakobsson & Eriksson 2000).

Germination and seedling establishment tend to be the most vulnerable stages in the life-cycle of plants because seedling mortality is often high (Harper 1997). However, germination ranks among the least studied aspects of bromeliad reproduction (Benzing 2000, but see Benzing 1978, 1981, Mondragón 2000). Studies on seedling survival report high mortality in the first year and survival increasing with plant size. Drought and/or branchfall were the most important causes of death (Benzing 1978, 1981, Hietz 1997). The implications of spatial heterogeneity on germination and survival are well known for other plants (e. g. Augspurger 1984, Kobe 1999, Daws *et al*. 2002, Beckage & Clark 2003, Castro *et al.* 2004), but have rarely been studied in epiphytes. Zotz (1997), studying the distribution of three epiphytic bromeliads, found no evidence for differential seedling mortality among growing sites and suggested that the spatial distributions of these epiphytes may be determined at the time of germination. Though survival was higher in the tree periphery and stem base than on intermediate branches, and on the sides rather than at the top or underside of branches in exposed seedlings of these bromeliads, this site-specific survival was not related to and could not help to explain the distribution of older individuals. (Zotz and Vollrath 2002)

A major constraint for the statistical analysis of seedling survival is that observations are often incomplete or censored, e.g., the birth or death of individuals is not observed, and data are generally non-normally distributed. Survival analysis is a powerful tool to analyse such data where the response variable is the time until occurrence of some event (Hosmer & Lemeshow 1999, Klein & Moeschberger 2003). The study of events involving an element of time has a long history in statistical research and practice, especially in clinical studies in medicine (e. g. Crowley & Hu 1977, Breslow & Day 1980, 1987, Goldberg *et al.* 1988). In ecological research, the use of survival analysis has only recently become more widespread, particularly the use of complex regression analysis of survival data (e. g. Greipsson & Davy 1994, Beckage & Clark 2003, Castro *et al.* 2004, Vange *et al.* 2004).

Microsite-specific germination or seedling survival could explain the observed distribution of individuals, but these are only two factors affecting the community structure of epiphytes. In this study germination and survival experiments with five epiphytic bromeliad species on branches in different canopy positions in a Mexican montane forest were conducted and in situ germination and seedling survival in relation to habitat characteristics were studied. Special attention was paid to differences in germination and seedling survival among canopy strata, branch inclinations, light and moisture conditions and to seasonal variations in mortality.

3.2 Materials and Methods

Study area and species

This study was conducted in a small forest reserve adjacent to the Instituto de Ecología, 2.5 km south of Xalapa, in central Veracruz, Mexico (19°31'N, 96°57'W), at 1350 m elevation. Average temperature is 19° C, and annual precipitation is 1500 mm, most of which falls in the June to October wet season. According to the Holdridge life-zone system (Holdridge 1967), the forest is at the transition between premontane and lower montane moist forest. In Mexico, it is commonly classified as 'bosque mesófilo de montaña' (mesophilous montane forest) following Rzedowski (1986). Descriptions of the forest structure are given by Williams-Linera (1997) and of the epiphyte community by Hietz and Hietz-Seifert (1995).

Germination and seedling survival were recorded in the epiphytic bromeliads *Catopsis sessiliflora* (Ruiz and Pav.) Mez, *Tillandsia deppeana* Steud., *T. juncea* (Ruiz and Pav.) Poir., *T. multicaulis* Steud. and *T. punctulata* Schltdl. and Cham. All *Tillandsia* juveniles and adults of *T. juncea* exhibit the atmospheric habit meaning that they possess narrow leaves that do not hold water but display confluent layers of absorbing trichomes. *C. sessiliflora* of all sizes and intermediate to adult *T. multicaulis* and *T. deppeana* have broad thin leaves that form tanks, and *T. punctulata* is an atmospheric-tank intermediate species.

Germination experiments

Seeds were collected in spring 2002 and stored dry at room temperature until used for the germination experiments. The viability of the seeds was not tested prior to exposure but to minimize possible maternal effects, the seeds of several plants were mixed. Of each study species, 20 randomly selected seeds were tied to 30 woody sticks (ca. 30 cm long, 5 cm diameter) with a thin thread. As bromeliad germination can be affected by substrate (Benzing 1978, Merwin *et al.* 2003), which is primarily a consequence of different water-holding capacities of the bark (Callaway *et al.* 2002), we used wood limbs of homogenous size and surface free from bryophytes and lichens. Lichens were removed because they can be allelopathic (Lawrey 1986) and also affect abiotic characteristics of the substrate (Hawkes & Menges 2003) by absorbing and retaining moisture (Riefner & Bowler 1995). The sticks were tied to branches on 11 trees, representing the range of canopy strata colonised by the species studied. According to their height of attachment relative to total tree height and distance to the trunk relative to the crown radius, branches were assigned to one of three canopy positions (trunk and inner crown, intermediate crown, outer crown). Canopy openness (using a convex spherical densiometer, Ben Meadows, Atlanta, GA), inclination, and circumference at the position of attachment were measured and percent bryophyte cover was estimated for each branch.

The sticks were exposed in May (*C. sessiliflora*), June (*T. deppeana*) and July (*T. juncea, T. multicaulis, T. punctulata*) 2002, at the times the seeds of the respective species are dispersed. Sticks were examined for seed germination and seedling survival approximately every tenth week after exposure until February 2003, and additionally at the end of the dry season in May 2003 and in February 2004.

Census of natural populations

In August 1999 and 2001, 186 branch sections between 10 and 250 cm long and distributed throughout the crowns of nine trees were selected and marked. The same branch parameters as for the germination experiments were measured and bryophyte and lichen cover were estimated. Natural germination and seedling survival of the study species were monitored twice a year until February 2003, and in February 2004. Fifty-eight seedlings of *C. sessiliflora* were recorded. As seedlings of *Tillandsia* species can not normally be determined to species level, the 368 seedlings were assigned to the group "*Tillandsia* juveniles". In cases where mass germination of seedlings was observed close to the fruiting mother plant, 388 seedlings of *T. deppeana* and 859 of *T. multicaulis* could be assigned to a species.

Statistical methods

Logistic regression was used to determine which of the following factors affect seed germination probabilities: species identity, position in the host tree, inclination, bryophyte cover, and canopy openness. Several models of different complexities were explored using the glm-function for binomial responses with logit-link in S-Plus (Anonymous 1999). The optimal model chosen was the one that minimised the Akaike information criterion (AIC = -2 log likelihood minus the number of parameters in model; Agresti 1990). To analyse time from exposure to germination and to seedling death, survival analysis was applied (Hosmer & Lemeshow 1999, Klein & Moeschberger 2003). Our survival data are right-censored because many individuals were still alive at the end of the observation, and interval-censored as survival time is only known to be between two observation times. The non-negative random variable T measures survival time. The survivor function S(t) measures the probability that an individual will survive beyond time t: S(t) = P[T≥t]

The Kaplan-Meier estimator S(t) was used to calculate nonparametric estimates of the survivor function:

$$\hat{S}_{(t)} = \prod_{j=1}^{s} \left(1 - \frac{d_j}{n_j} \right)$$

where d_j is the number of individuals that experienced the event in a given interval and n_j is the number at risk. Survival curves are monotone non-increasing step functions equal to one at time zero, and zero as time approaches infinity. Statistical differences among survival curves were calculated using the log-rank test. Because this test cannot be calculated for interval-censored data, the mid-point of each interval was used in this case.

To account for explanatory variables, parametric regression models of the form $f(T) = \beta_0 + \beta^t x + \sigma\varepsilon$ were fitted to the data using the censorReg function in S-plus 2000 (Anonymous, 1999), where $f()$ is a model-dependent link function, β_0 is the intercept, β^t is a transposed vector of regression coefficients, x is a vector of explanatory variables, σ is the scale parameter and ε is the model-dependent error distribution. Explanatory variables were position, inclination, bryophyte cover and canopy openness. Covariates were included in the models following the purposeful selection procedure proposed by Hosmer and Lemeshow (1999). All covariates significant at the 20 percent level in the bivariate analysis were included in an initial multivariable model. The values of the Wald tests were used to identify non-significant covariates which were removed one by one from the model. Following the fit of the reduced model, we assessed if removal of a covariate produced a change of more than 20 percent in the coefficients of the variables remaining in the model. If the variable excluded was an important confounder it was added back into the model. This process continued until no more covariates could be deleted from the model. In the next step, all two-way interactions from the main terms were added to the main effects model. All interactions significant at the 5 percent level (likelihood ratio test) were then added jointly to the model. Interactions that were still significant remained in the model. A frailty term was incorporated to account for unmeasured "random" effects of the branches.

Models with different error distributions were explored (normal and logistic with an identity link; Weibull, exponential, extreme, logistic, log-logistic and log-normal which have a natural log link) and compared using the Akaike information criterion statistic. The model with the smallest AIC value was chosen as the best fitting final model. The coefficient values, their standard error, 95 % confidence interval, significance and time ratio (Hosmer & Lemeshow 1999) are reported. The time ratio (e^β) reflects how a change in covariate values changes survival time (e. g. when time ratio for a dichotomous covariate with $x = 0$ and $x = 1$ is 2.0, the interpretation is that survival times of subjects with $x = 1$ is twice that of individuals with $x = 0$).

Daily mortality in the rainy season 2002 (June to October), in winter 2002/2003 (November to February), in the dry season 2003 (March to May) and between May 2003 and February 2004 (rainy season plus winter) was calculated as

$m = 1 - (N_1/N_0)^{1/t}$

where N_0 and N_1 are seedling counts at the beginning and end of the census interval t (Sheil et al. 1995).

All statistical analyses were calculated using S-plus 2000 (MathSoft Inc. 1988-1999).

3.3 Results

Germination experiments

The best logistic regression model was the null model without explaining variables, thus germination probabilities were not affected by any of the observed parameters. Out of 3000 seeds 524 germinated, most of them (65.8 %) within the first ten weeks after exposure (Table 3.1). Most of the seeds of *T. multicaulis* failed to germinate. The median time to germination was 37 days in *T. punctulata* and *T. juncea*, 40 days in *C. sessiliflora* and 62 days in *T. deppeana*. According to the log-gaussian parametric survival model, time to germination was species specific and affected by position in the crown (Table 3.2A), with seeds exposed in the outer crown taking longer to germinate. The median time to germination in the inner, intermediate and outer crown was 37, 40 and 44 days, respectively. *Tillandsia juncea* seeds germinated sooner on steeply inclined branches, in the other species the interaction with branch inclination was not significant (Table 3.2A).

Table 3.1: Percentage of germinated seeds within 10, 20, and 30 weeks and percentage of seedlings that survived 7, 12 and 20 months in the experiments.

	% germinated within			% seedlings alive		
	10 weeks	20 weeks	30 weeks	ca. 7 months	ca. 12 months	ca. 20 months
Catopsis sessiliflora	16.5	18.5	18.8	14.0	7.0	3.5
Tillandsia deppeana	4.2	6.4	7.2	27.9	11.6	4.7
T. juncea	31.5	33.7	33.7	24.8	17.8	9.4
T. multicaulis	0.2	0.2	0.2	0.0	0.0	0.0
T. punctulata	25.8	27.1	27.3	16.5	10.4	4.9
total	15.7	17.2	17.5	20.0	12.6	6.3

Seedling survival was similar among species (Figure 3.1A, Table 3.3). About two thirds of newly germinated seedlings were dead in the following census (*T. juncea*: 56.9 %, *T. punctulata*: 67.1 %,

T. deppeana 72.1 %, *C. sessiliflora* 77.2 %). Less than 20 % of the seedlings or 3 % or less of the seeds were still alive about one year after exposure (Table 3.1). The most parsimonious parametric survival model was the log-logistic model (Table 3.2B). Seedling survival time was influenced by position in the crown and canopy openness (Table 3.2B, Figure 3.2A): in outer and intermediate crown positions, survival time was longer. An increase in canopy openness of 10 % increased seedling survival time by the factor 1.11. Compared to the reference species *C. sessiliflora*, median survival time was significantly higher for *T. deppeana* seedlings (Table 3.2B).

Figure 3.1: Seedling survival in epiphyte species based on Kaplan-Meier estimates in experiments (A) and in census populations (B). Differences in survival curves are significant (log-rank test, P<0.001).
Cs = *Catopsis sessiliflora*, Td = *Tillandsia deppeana*, Tj = *T. juncea*, T.juv. = *Tillandsia* juveniles not identified to species level, Tm = *T. multicaulis*, Tp = *T. punctulata*.

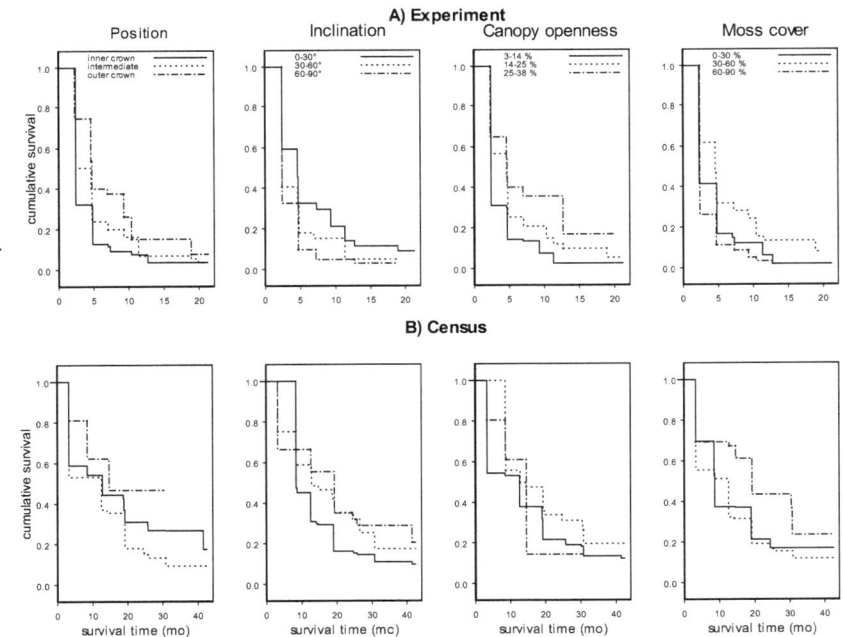

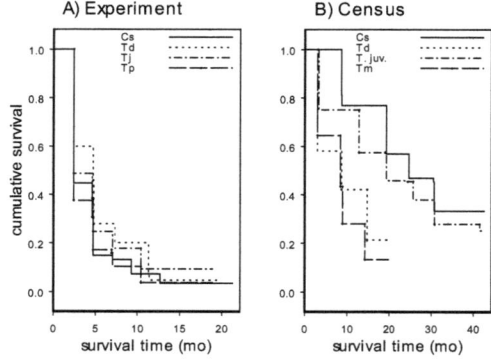

Figure 3.2: The effect of the branch parameters position, branch inclination, canopy openness and bryophyte cover on seedling survival in experiments (A) and census populations (B). Survival functions were based on Kaplan-Meier estimates. Differences in survival curves are significant (log-rank test, P<0.001).

Table 3.3: Median survival time in experimental and census populations. *Tillandsia* juveniles could not be identified to species level.

	Median survival time (days)	
	Experiments	Census
Catopsis sessiliflora	132	778
Tillandsia deppeana	143	354
T. juncea	137	-
T. multicaulis	-	256
T. punctulata	125	-
Tillandsia juveniles	-	558

Table 3.4: Per cent mortality per day on shady branches, where canopy openness is below the mean of all branches (< 17 %), and sunny branches (canopy openness >= 17 %) in different seasons. Differences between branch types are not significant (Mann-Whitney U-test, P>0.05).

Species	branch	% mortality d^{-1} (n)			
		rainy season 2002	winter 2002/03	dry season 2003	rainy season + winter 2003/04
Catopsis sessiliflora					
	shady	1.35 (54)	0.00 (14)	0.89 (15)	0.11 (9)
	sunny	1.55 (45)	0.49 (14)	0.18 (13)	0.35 (12)
Tillandsia deppeana					
	shady	2.27 (20)	1.61 (14)	0.79 (8)	0.42 (3)
	sunny	0.73 (5)	0.90 (6)	0.56 (4)	0.27 (2)
T. juncea					
	shady	1.67 (116)	1.04 (39)	0.37 (19)	0.37 (13)
	sunny	0.65 (73)	0.83 (55)	0.29 (31)	0.19 (23)
T. punctulata					
	shady	2.01 (110)	1.34 (33)	0.61 (13)	0.33 (7)
	sunny	0.94 (45)	0.88 (24)	0.33 (14)	0.47 (10)

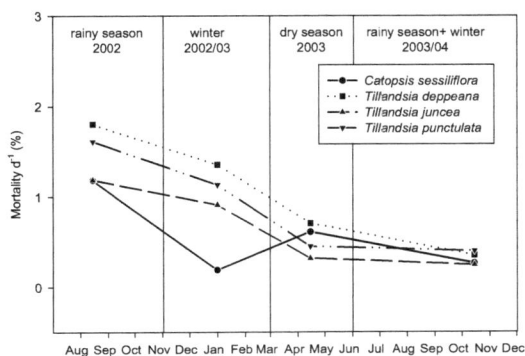

Figure 3.3: Per cent mortality per day in epiphyte species in different seasons.

Daily mortality decreased with increasing seedling age (Figure 3.3). The *Tillandsia* species showed an almost parallel monotonic decline in mortality. In contrast, daily mortality in *C. sessiliflora* seedlings increased dramatically in the dry season. For every species except *C. sessiliflora* daily mortality was higher on shaded than on sun-exposed branches (Table 3.4).

Census of natural populations

Seedlings that had germinated naturally on branches survived longer than those attached to the sticks (Figure 3.4), and survival time differed among species (Figure 3.1B, Table 3.3). Pooling all species, survival was higher on steep branches and when bryophyte cover was high (Figure 3.2B, Table 3.2C). A log-gaussian survival model revealed that the interaction between species and bryophyte cover and between species and inclination significantly influenced survival time (Table 3.2C). In *C. sessiliflora* and *T. deppeana*, median survival time increased with bryophyte cover (though not significantly in the latter), whereas in *T. multicaulis* survival time was shorter on branches with strong bryophyte cover. Steeper branches were favourable for the seedling survival of *T. deppeana* and *T. multicaulis*, but not for *C. sessiliflora*.

Table 3.2: Parametric survival models for interval and right-censored data of A) germination, B) seedling death in experiments and C) seedling death in the census populations. The coefficient values (β), their standard error, Z and p-value of the Wald test and time ratio are listed. The time ratio reflects the amount by which median survival time is changed by a factor (e^{β}) or a ten-unit increase ($e^{\beta*10}$) in a continuous covariate. The frailty term accounts for unmeasured random effects of the branches. Factor variables are split into 0-1 coded dummy variables according to their number of factor levels. The first (= reference) level has a coefficient-value of zero and time ratio is one. The reference levels in the variable species are *C. sessiliflora* in the experiments, and unidentified *Tillandsia* juveniles in the census survival model, respectively. In the covariate position, "inner crown" is the reference level.

	Coefficient	SE	Z	p	time ratio
A) Time to Germination (log-gaussian, n = 523)					
(Intercept)	0.441	0.138	3.20	0.001	
Species: Td	0.287	0.078	3.68	0.000	1.33
Species: Tj	-0.307	0.073	-4.20	0.000	0.74
Species: Tp	-0.065	0.034	-1.89	0.059	0.94
Position: intermediate	0.154	0.068	2.28	0.023	1.17
Position: outer crown	0.111	0.041	2.69	0.007	1.12
Inclination	-0.003	0.002	-1.35	0.176	1.00
frailty(branch)	-0.010	0.004	-2.74	0.006	
Td*Inclination	0.002	0.003	0.78	0.438	1.02
Tj*Inclination	-0.006	0.002	-3.17	0.002	0.94
Tp*Inclination	0.001	0.001	0.75	0.457	1.01
B) Seedling survival experiment (log-logistic, n = 523)					
(Intercept)	1.656	0.118	13.98	0.000	
Species: Td	0.166	0.049	3.41	0.001	1.18
Species: Tj	0.014	0.021	0.69	0.494	1.01
Species: Tp	-0.018	0.014	-1.29	0.197	0.98
Position: intermediate	0.048	0.030	1.59	0.111	1.05
Position: outer crown	0.069	0.025	2.82	0.005	1.07
Inclination	-0.002	0.001	-2.01	0.045	0.98
Bryophyte	-0.001	0.001	-1.11	0.267	0.99
Canopy openness	0.010	0.004	2.52	0.012	1.11
frailty(branch)	-0.004	0.002	-2.38	0.017	
C) Seedling survival census (log-gaussian, n = 1673)					
(Intercept)	2.385	0.105	22.73	0.000	
Species: Cs	-0.405	0.154	-2.62	0.009	0.67
Species: Td	-0.228	0.076	-3.01	0.003	0.80
Species: Tm	0.092	0.035	2.65	0.008	1.10
Inclination	-0.002	0.001	-1.20	0.230	0.98
Bryophyte	0.012	0.002	7.00	0.000	1.12
frailty(branch)	-0.006	0.001	-6.64	0.000	
Cs*Bryophyte	0.010	0.003	3.37	0.001	1.10
Td*Bryophyte	0.001	0.001	0.85	0.393	1.01
Tm*Bryophyte	-0.005	0.001	-7.05	0.000	0.95
Cs*Inclination	-0.007	0.003	-2.64	0.008	0.93
Td*Inclination	0.002	0.001	1.65	0.099	1.02
Tm*Inclination	0.002	0.001	3.93	0.000	1.02

Cs = *Catopsis sessiliflora*, Td = *Tillandsia deppeana*, Tj = *T. juncea*, Tm = *T. multicaulis*, Tp = *T. punctulata*

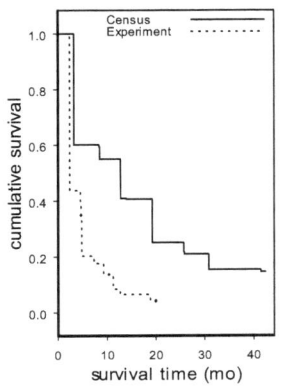

Figure 3.4: Survival of all seedlings based on Kaplan-Meier estimates under experimental and natural (census) conditions. Differences in survival curves are significant (log-rank test, P<0.001).

3.4 Discussion

Germination experiments

Seeds were exposed as capsules of all five species opened during the dry season or at the beginning of the rainy season (unpubl. data). Many anemochorous epiphytes disperse their seeds during the dry season when deciduous trees lose their foliage and movement through canopies are not hindered by the leaves of trees (Madison 1977). However, in our forest most of the dominant trees shed their foliage in winter, and by the beginning of the dry season new leaves are expanding (Williams-Linera & Tolome 1996). The fact that precipitation tends to mat the comas of these seeds also favours dry season dispersal as does dry air itself (Bonn & Poschlod 1998).

The failure of the planted seeds of *Tillandsia multicaulis* to germinate is surprising considering its abundance at the study site. It is possible that the seeds, collected from seven plants, were not viable or that the species has particular germination requirements, though the latter appears unlikely as this species is found growing on a wide range of branch types and positions. In the other species, germination ranged from 7.2 to 33.7 %. These values may be underestimates as it is possible that some seeds germinated and died between two observation intervals, but they are still in the upper range observed in other bromeliad or epiphyte species. Benzing (1978, 1981) obtained 6 - 35 % germination of *Tillandsia paucifolia* after 14 weeks in greenhouse experiments with regular misting but less than 4 % in the field. In a Mexican dry forest planted *Tillandsia brachycaulos* seeds germinated at 2.4, 3.1 and 4.2 % in three consecutive years (Mondragón 2000). Nearly 100 % of the seeds of the epiphytic orchid *Laelia speciosa* (Hernández-Apolinar 1992) germinated in vitro, but results in the field were far below 1 % ($4.8 \times 10^{-5} - 2.2 \times 10^{-4}$). Between 9 and 41 % of the seeds of hemiepiphytic *Ficus stupenda* germinated on different substrate types in Borneo (Laman 1995).

Whether germination occurred or not was not influenced by any of the investigated parameters. However, lower germination rates in situ compared to in under laboratory conditions in several *Tillandsia* species (Benzing 1978, 1981, García-Franco 1990, Mondragón 2000) show that some factors must be responsible for sub-optimal germination conditions in the field. About 90 % of germination events were observed within ten weeks of exposure. A single seed germinated after 30 weeks and the remaining seeds appeared to be dead or never were viable. Our findings thus support the notion that epiphyte seeds are non-dormant and do not build seed-banks (Benzing 1990). In laboratory experiments all of the seeds of *Tillandsia deppeana* that germinated did so within seven days (García-Franco 1990). Seeds of all species germinated earlier in the less exposed zones of the canopy where branches are moister, suggesting that insufficient humidity delays germination in the field.

Seedling establishment was low, only one eighth of them survived after one year and only half of these were still living after 20 months. The early stages of the cycle are the most vulnerable in epiphytes as in most other vascular plants (Silvertown & Lovett-Doust 1993). For instance, first year survival in seedlings of *Tillandsia paucifolia* was only between 0.46 and 3.35 % in seedling cohorts of four consecutive years (Benzing 2000). Mondragón *et al.* (2004) report annual seedling mortalities between 21 and 71 % in three years for *Tillandsia brachycaulos*. The low survival rates contrast with a previous study that found survival probabilities of about 70% for *Catopsis* and *Tillandsia* plants < 2 cm length at the same forest site (Hietz 1997). However, that study was based on the analysis of photographs taken annually, where all individuals that germinated and died between two observations were left out, and these seedlings were larger and probably less vulnerable than the seedlings described in this study.

Survival times were longer in the outer canopy and in locations where canopy openness was high. In the atmospheric seedlings of *Tillandsia* spp, mortality steadily decreased with age, but in the more mesomorphic *C. sessiliflora* mortality increased in the dry season. By contrast, dry season mortality was lowest in *T. juncea* seedlings, the only species utilising CAM (Hietz *et al.* 2002).

Furthermore, *Tillandsia* seedlings mortality was lower on sunny branches during all seasons. This is surprising because drought constitutes the major threat to early juveniles because they desiccate faster than adults due to their less favourable surface-volume ratio and small size (Benzing 1990, Hietz 1997, Schmidt & Zotz 2001, Zotz *et al.* 2001, Mondragón *et al.* 2004). Seedlings of the soft-leafed, tank-forming Tillandsioideae are equipped with denser layers of insulating and absorbing trichomes than the later life stages (Madison 1977, Adams & Craig 1986). The trichomes aid in water uptake and reduce water loss but hinder photosynthesis (Adams & Craig 1986) when the wet leaf surface impedes gas exchange (Benzing *et al.* 1978). Thus, atmospheric species as well as atmospheric seedlings of species that as adults have broad leaves with only scattered trichomes on their surface may be favoured by exposed conditions, not only because they require more light, but also because they are damaged by excessive humidity.

Census of natural population

The longer survival times among the natural populations compared to those exposed for the experiments may in part be a consequence of longer observation intervals, which were mostly six months for the census but two months for the experimental population. Thus, any seed that germinated and died between two observations was not recorded. High early stage mortality, only 14 – 28 % of the experimental seedlings survived for 7 months, largely explains the difference in survival between natural and experimental seedlings (Figure 3.4). Additionally, the sticks with rather smooth bark and few cryptogams used in the experiments may also have been less suitable than the live branches that support the census population.

Most seedlings were anchored near what could have been their parents. Spatial distribution of seeds is usually clumped because few of them travel more than short distances from their parents (Silvertown & Lovett-Doust 1993). About half of the seeds of wind-dispersed *Tillandsia deppeana* remained near the mother plant, and for more mobile seeds the probability of reaching a branch decreased with the distance travelled (García-Franco & Rico-Gray 1998). Of 8500 seeds used in their experiments only 2 % were recovered from one of the six downwind traps that covered in total 3.4 m². Seventy-two seeds per square metre were captured at a distance of 8.8 m from the experimental "mother plant", 54 at 14.4 m, and 44 at 28 m of distance, respectively (García-Franco & Rico-Gray 1998).

On branches in situ, which had a more diverse bark structure and cryptogam cover than the experimental sticks, bryophyte cover was more important for survival than light. Survival times of seedlings of *C. sessiliflora* and *T. deppeana* increased with bryophyte cover. These species occur in more exposed canopy strata (Hietz *et al.* 2002) where the humidity and possibly protection against excess radiation provided by cryptogams may favour seedling success. On the other hand, on the thicker and shadier branches preferred by *T. multicaulis* (Hietz *et al.* 2002), bryophyte cover negatively affected seedling survival, probably because of excessive humidity or shading of seedlings.

Conclusions

The heterogeneity of the habitat in a sampled forest canopy, which presumably influences the distribution of epiphytes growing there, affected seedling success, but not germination rates. Given the scarcity of safe sites in epiphytic habitats and the high initial mortality, germination and early survival can be bottlenecks determining epiphyte population sizes and growth rates, but further research is needed to confirm this. Because epiphyte seedlings are sensitive to differences in microclimate and their mortality is typically high, we expect that global climate change will affect epiphytes primarily during their seedling stage. If tropical wet mountains get drier, as has been predicted and in part already observed (Pounds *et al.* 1999), epiphytes may be the most adversely impacted group (Benzing 1998, Foster 2001). We expect many epiphytes to survive only in relatively protected positions within the habitats they occupy now, and to possibly colonize higher elevation sites if such forests are present in an area and if migration is fast enough.

3.5 Acknowledgements

We thank Leticia Cruz Paredes and Angélica Jiménez Aguilar for help in the field. We are grateful to the botanical garden Francisco Clavijero and the Instituto de Ecología in Xalapa and in particular to José García Franco for general support. Barbara Holzinger and two anonymous reviewers provided helpful comments on an earlier draft of the manuscript. This research was funded by the Austrian Science Fund (FWF grant number P14775) to P.H.

3.6 Literature

Adams W. W. & Craig M. E. 1986. Physiological consequences of changes in life form of the the Mexican epiphyte *Tillandsia deppeana* (Bromeliaceae). *Oecologia* **70**: 298-304.

Agresti A. 1990. *Categorical data analysis.* New York: Wiley.

Anonymous. 1999. *S-PLUS 2000. Guide to Statistics, Volume 2.* Seattle: Data Analysis Products Division, MathSoft.

Augspurger C. K. 1984. Light requirements of neotropical tree seedlings —a comparative study of growth and survival. *Journal of Ecology* **72**: 777–795.

Beckage B. & Clark J. S. 2003. Seedling survival and growth of three forest tree species: the role of spatial heterogeneity. *Ecology* **84**: 1849-1861.

Benzing D. H. 1978. Germination and early establishment of *Tillandsia circinnata* Schlecht. (Bromeliaceae) on some of its hosts and other supports in southern Florida. *Selbyana* **5**: 95-106.

Benzing D. H. 1981. The population dynamics of *Tillandsia circinnata* (Bromeliaceae): cypress crown colonies in southern Florida. *Selbyana* **5**: 256-263.

Benzing D. H. 1990. *Vascular Epiphytes. General Biology And Related Biota.* Cambridge: Cambridge University Press.

Benzing D. H. 1998. Vulnerabilities of tropical forests to climate change: the significance of resident epiphytes. *Climate Change* **39**: 519–5440.

Benzing D. H. 2000. *Bromeliaceae: Profile of an Adaptive Radiation.* Cambridge: Cambridge University Press.

Benzing D. H, Seemann J. & Renfrow A. 1978. The foliar epidermis in Tillandsioideae (Bromeliaceae) and its role in habitat selection. *American Journal of Botany* **65**: 359-365.

Bonn S. & Poschlod P. 1998. *Ausbreitungsbiologie der Pflanzen Mitteleuropas.* Wiesbaden: Queller, Meyer Verlag.

Breslow N. E. & Day N. E. 1980. *Statistical methods in cancer research, Volume I: The analysis of case-control studies.* Oxford: Oxford University Press.

Breslow N. E. & Day N. E. 1987. *Statistical methods in cancer research. Volume II: The design and analysis of cohort studies.* Oxford: Oxford University Press.

Callaway R. M., Reinhart K. O., Moore G. W., Moore D. J. & Pennings S. C. 2002. Epiphyte host preferences and host traits: Mechanisms for species-specific interactions. *Oecologia* **132**: 221-230.

Castro J., Zamora R., Hódar J. A. & Gómez J. M. 2004. Seedling establishment of a boreal tree species (*Pinus sylvestris*) at its southernmost distribution limit: consequences of being in a marginal Mediterranean habitat. *Journal of Ecology* **92**: 266-277.

Crowley J. & Hu M. 1977. Covariance analysis of heart transplant survival data. *Journal of American Statistical Association* **78**: 27-36.

Daws M. I., Burslem D. F., Crabtree L. M., Kirkman P., Mullins C. E. & Dalling J. W. 2002. Differences in seed germination responses may promote coexistence of four sympatric *Piper* species. *Functional Ecology* **16**: 258-267.

Foster P. 2001. The potential negative impacts of global climate change on tropical montane cloud forests. *Earth-Science Reviews* **55**: 73-106.

García-Franco J. G. 1990. *Biología reproductiva de Tillandsia deppeana Strudel (Bromeliaceae)*. Master Thesis, Universidad Nacional Autónoma de Mexico, Mexico.

García-Franco J. G. & Rico-Gray V. 1998. Experiments on seed dispersal and deposition pattern of epiphytes - the case of *Tillandsia deppeana* Steudel (Bromeliaceae). *Phytologia*, **65**: 73-78.

Goldberg R. J., Gore J. M., Alpert J. S. & Dalen J. E. 1988. Incidence and case fatality rates of acute myocardial infarction (1975-1984): The Worcester Heart Attack Study. *American Heart Journal* **115**: 761-767.

Greipsson S. & Davy A. J. 1994. Germination of *Leymus arenarius* and its significance for land reclamation in Iceland. *Annals of Botany* **73**: 393-401.

Hanski I. & Gilpin M. E. 1997. *Metapopulation biology. Ecology, genetics, and evolution.* San Diego: Academic Press.

Harper J. L. 1977. *Population biology of plants*. New York: Academic Press.

Hawkes C. V. & Menges E. S. 2003. Effects of lichens on seedling emergence in a xeric Florida shrubland. *Southeastern Naturalist* **2**: 223-234.

Hernández-Apolinar M. 1992. *Dinámica poblacional de* Laelia speciosa *(H. B. K.) Schltr. (Orchidaceae)*. Tesis de Licenciatura, Universidad Autónoma de México, Mexico.

Hietz P. 1997. Population dynamics of epiphytes in a Mexican humid montane forest. *Journal of Ecology* **85**: 767-775.

Hietz P., Ausserer J. & Schindler G. 2002. Growth, maturation and survival of epiphytic bromeliads in a Mexican humid montane forest. *Journal of Tropical Ecology* **18**: 177-191.

Hietz P. & Hietz-Seifert U. 1995. Structure and ecology of epiphyte communities of a cloud forest in central Veracruz, Mexico. *Journal of Vegetation Science* **6**: 719-728.

Holdridge L. R. 1967. *Life zone ecology*. San José, Costa Rica: Tropical Science Center.

Hosmer D. W. Jr. & Lemeshow S. 1999. *Applied survival analysis. Regression modeling of time to event data*. New York: John Wiley, Sons.

Howe H. F. & Smallwood J. 1982. Ecology of seed dispersal. *Annual Review of Ecology and Systematics* **13**: 201-228.

Jakobsson A. & Eriksson O. 2000. A comparative study of seed number, seed size, seedling size and recruitment in grassland plants. *Oikos* **88**: 494-502.

Klein J. P. & Moeschberger M. L. 2003. *Survival analysis. Techniques for censored and truncated data, 2nd edition*. New York: Springer.

Kobe R. K. 1999. Light gradient partitioning among tropical tree species through differential seedling mortality and growth. *Ecology* **80**: 187–201.

Laman T. G. 1995. *Ficus stupenda* germination and seedling establishment in a Bornean rain forest canopy. *Ecology* **76**: 2617-2626.

Lawrey J. D. 1986. Biological role of lichen substances. - The Bryologist, 89: 111-122.

Madison M. 1977. Vascular epiphytes: their systematic occurrence and salient features. *Selbyana* **2**: 1-13.

Merwin M. C., Rentmeester S. A., Nadkarni N. M. 2003. The influence of host tree species on the distribution of epiphytic bromeliads in experimental monospecific plantations, La Selva, Costa Rica. *Biotropica* **35**: 37-47.

Mondragón D. 2000. *Dinámica poblacional de Tillandsia brachycaulos Schltdl. en el parque nacional de Dzibilchaltún, Yuc.* PhD thesis. Centro de Investigación Científica de Yucatán, México.

Mondragón D., Durán R., Ramírez I. & Valverde T. 2004. Temporal variation in the demography of the clonal epiphyte *Tillandsia brachycaulos* (Bromeliaceae) in the Yucatán Peninsula, Mexico. *Journal of Tropical Ecology* **20**: 189-200.

Pounds J. A., Fogden M. P. & Campbell J. H. 1999. Biological response to climate change on a tropical mountain. *Nature* **398**: 611-615.

Riefner R. E. Jr. & Bowler P. A. 1995. Cushion-like fruticose lichens as *Dudleya* seed traps and nurseries in coastal communities. *Madroño* **42**: 81-82.

Rzedowski J. 1986. *Vegetación de México, 3rd edition*. Mexico: Editorial Limusa.

Schmidt G. & Zotz G. 2001. Ecophysiological consequences of differences in plant size: in situ carbon gain and water relations of the epiphytic bromeliad, *Vriesea sanguinolenta*. *Plant, Cell and Environment* **24**: 101-111.

Sheil D., Burslem D. F. & Alder D. 1995. The interpretation and misinterpretation of mortality rate measures. *Journal of Ecology* **83**: 331-333.

Silvertown J. W. & Lovett Doust J. 1993. *Introduction to Plant Population Biology*. Oxford: Blackwell Science.

van der Pijl L. 1972. *Principles of dispersal in higher plants, 2nd edition*. Berlin: Springer.

Vange V., Vandvik V. & Heuch I. 2004. Does seed mass and family influence germination and juvenile performance in the *Knautia arvensis*? A study using failure-time methods. *Acta Oecologia* **25**: 169-178.

Williams-Linera G. & Tolome J. 1996. Litterfall, temperate and tropical dominant trees, and climate in a Mexican lower montane forest. *Biotropica* **28**: 649-656.

Williams-Linera G. 1997. Phenology of deciduous and broadleaved-evergreen tree species in a Mexican tropical lower montane forest. *Global Ecology and Biogeography Letters* **6**: 115-127.

Zotz G. 1997. Substrate use of three epiphytic bromeliads. *Ecography* **20**: 264-270.

Zotz G., Hietz P. & Schmidt G. 2001. Small plants, large plants: the importance of plant size for the physiological ecology of vascular epiphytes. *Journal of Experimental Botany* **52**, 2051-2056.

Zotz G. & Vollrath B. 2002. Substrate preferences of epiphytic bromeliads: An experimental approach. *Acta Oecologica* **23**: 99-102.

4 Breeding systems, fruit set, and flowering phenology of epiphytic bromeliads and orchids in a Mexican humid montane forest

Peter Hietz, Manuela Winkler, Leticia Cruz-Paredes & Angélica Jímenez-Aguilar

published in Selbyana (2006) 27: 156 – 164 (reprinted with permission)

Abstract

Epiphyte pollination is constrained by a stressful habitat that limits the amount of resources to be invested in pollinator attraction. Other constraints are the difficulty of locating conspecific individuals in sometimes highly dispersed populations within the canopy and the ephemeral substrate where branch failure may cut short the time available for successful reproduction. The authors conducted pollination experiments on the relationships of breeding systems, pollination success, flowering phenology, and microhabitat preference in epiphytic orchids and bromeliads in a Mexican humid montane forest. Phenology and fruit set also were observed. The breeding systems ranged from dioecious (*Catopsis sessiliflora*) to largely or entirely self-incompatible and outcrossing (*Tillandsia multicaulis*, *T. punctulata*, and *Lycaste aromatica*) to partly or mainly self-pollinating (*T. juncea*, *Jacquiniella teretifolia*, and probably *J. leucomelana*). Fruit set in the field was highest in the orchid *Jacquiniella teretifolia* (76–88%) and in the bromeliad *Catopsis sessiliflora* (71%), both of which grow preferentially on more exposed branches. Ranked next were monocarpic *Tillandsia deppeana* (60%) and xeric *T. juncea* (60%). Fruit set was lower in *J. leucomelana* (29–40%), *T. multicaulis* (41%), and *T. punctulata* (25%) and lowest in long-lived *L. aromatica* (8–11%), plants of which grow mostly on stable branches. The trend for selfing and/or higher fruits sets found in species growing on more ephemeral branches or adapted to more resource-poor conditions suggests that epiphyte pollination reflects adaptations to the diversity of canopy microsites.

Key words: breeding system, Bromeliaceae, epiphyte, fruit set, Mexico, Orchidaceae

4.1 Introduction

In zoophilous plants, investment in pollinator attraction increases the probability of being visited, but current investment in reproduction may reduce future growth or reproduction. Life-history theory suggests that plants invest resources to maximize reproductive output, and that evolution selects plants not limited by pollinators but by resources invested either in attraction or in seed production (Janzen 1977, Ashman *et al.* 2004). Theory also suggests that plant pollination systems optimize the balance between the chance of an ovule being fertilized by cross-pollination and the genetic disadvantage of inbreeding-depression through self-pollination (Tanaka 1997, Johnsen *et al.* 2003). Though some species effectively exclude selfing by being dioecious or self-incompatible, others take no risks and self-pollinate, in which case they spare resources for attracting pollinators. Inbreeding depression is measured as reduced fruit set or reduced seed or embryo numbers in cross-pollinated versus self-pollinated plants. It is stronger in normally outcrossing than in selfing species, suggesting that the main cause of inbreeding depression (deleterious alleles) has been eliminated or reduced in the evolution of selfing species (Tremblay *et al.* 2005). Thus self-pollination ought to provide a relative advantage where pollinators are scarce or attracting them is too costly or in circumstances where reproduction is severely time-limited. The time-limitation hypothesis has been used to explain why most self-pollinating plants are annuals (Aarssen 2000) and why selfers are over-represented among annuals in particularly time-limited habitats (Snell & Aarssen 2005).

Tropical epiphytes are not annuals but, in most cases, polycarpic perennials that live in a habitat with severe resource limitation. In addition to the mostly small to moderate size of epiphytes, resources likely limit the potential for pollinator attraction (Ackerman 1986, Benzing 1990). Epiphytes, however, also live in an ephemeral habitat, where exfoliating bark, breaking branches, and falling trees impose limits on life expectancy. Epiphytes that fall to the ground usually die, and branch instability is known to be a factor of mortality and population growth (Hietz 1997, Zotz 1998, Hietz *et al.* 2001, Zotz & Schmidt 2006). With epiphyte diversity in humid tropical forests high and conspecific individuals often at a distance, pollinator specificity is significant and reflected in a high diversity of floral structures, as illustrated in the largest epiphytic family, the orchids (Ackerman 1986).

In plants where fruit maturation takes a relatively long time, as in most epiphytes, fruit set (proportion of developing fruits to total number of flowers present) is a useful and easy-to-obtain measure of pollination success (Neiland & Wilcock 1998). A higher fruit set does not necessarily result in an increased reproductive output, however, because the investment in more fruits may lead to fewer seeds or reduced vegetative and reproductive growth in the future (Montalvo & Ackerman 1987, Zimmerman & Aide 1989, Ackerman & Montalvo 1990, Bartareau 1995, Melendez-Ackerman et al. 2000). Detailed demographic studies with experimental variation of fruit sets show that the low fruit set found in many orchids does indeed indicate pollinator limitation (Calvo & Horvitz 1990, Calvo 1993). Orchids appear to be often pollinator-limited (Neiland & Wilcock 1998, Tremblay et al. 2005) and Ashman et al. (2004) provide a theoretical framework for pollen limitation. Recent reviews report significantly higher natural fruit set in rewarding than in deceptive orchids, percentages about twice as high in temperate compared to tropical orchids (Tremblay et al. 2005), and no difference between epiphytic and terrestrial species (Neiland & Wilcock 1998).

While many tropical woody plants are self-incompatible (Bawa 1974, Bullock 1985, Kress & Beach 1994, Ward et al. 2005), all 22 epiphytes studied in a Costa Rican cloud forest were self-compatible (Lumer 1980, Bush & Beach 1995). Most orchids, including epiphytes, are also self-compatible (Dressler 1981, Tremblay et al. 2005), and a number are self-pollinating (Catling 1990). Self-incompatibility, however, appears to be common in some groups such as *Pleurothallis* (Borba et al. 2001) and Epidendroideae (Tremblay et al. 2005). Self-compatibility also was reported for at least 20 out of 35 Brazilian bromeliads (Martinelli 1994 cited in Benzing 2000) and autogamy for 8 out of 188 bromeliads from the Bolivian Andes (Kessler & Krömer 2000), although the latter data were largely inferred from pollination syndromes and not verified by experiments. Since epiphytes live in a resource-poor and more or less ephemeral habitat, which limits resources and time for reproduction, selection may favor pollination systems that enable successful reproduction in the absence of either a pollinator or a nearby conspecific. In epiphytes from xeric and/or very ephemeral habitats, where plant size is often reduced and the time for successful reproduction is short, the pressure for autogamy or geitonogamy may be particularly high (Benzing 1978, Gilmartin & Brown 1985). For the present study, breeding systems and natural fruit set of eight common epiphytic species in a Mexican humid montane forest were analyzed, looking for relationships between selfing, self-compatibility, phenology, fruit set, and the preferred microhabitat of a species. We hypothesized that mechanisms increasing reproductive output with less investment in pollinator attraction will be more common in the more xeric species and in species preferring thin and short-lived branches.

4.2 Methods

The study was conducted in the Botanical Garden Clavijero and a small forest reserve adjacent to the Instituto de Ecología, 2.5 km south of Xalapa, in central Veracruz, Mexico (19°31'N, 96°57'W) at ca. 1350 m. Average temperature is 19°C, and annual precipitation is 1500 mm, most of which falls in the wet season between June and October. According to the Holdridge life-zone system (Holdridge 1967), the forest is at the transition between premontane and lower montane moist forests. In Mexico, it is commonly classified as 'bosque mesófilo de montaña' (mesophilous montane forest; Rzedowski 1986). Descriptions of the forest structure are given by Williams-Linera (1997) and of the epiphyte community by Hietz & Hietz-Seifert (1995).

Three orchid and five bromeliad species were studied (Table 4.1). *Jacquiniella teretifolia* (Sw.) Britton & P.Wilson has caespitose erect stems with distichous, almost terete leaves and yellowish-green flowers, ca. 2 cm long, without distinct fragrance. *Jacquiniella leucomelana* (Rchb. f.) Schltr. is of similar habit, but stems and leaves are usually less than half as large and flowers measure only 3 mm. *Lycaste aromatica* (Graham ex Hook.) Lindl. has thin, drought-deciduous leaves at the top of a broad pseudobulb, with dark-yellow flowers arising from the base of the pseudobulb. The flowers emit a strong smell of cinnamon and are pollinated by euglossine bees (Dressler 1968). *Tillandsia deppeana* Steud. produces a large impounding rosette and a pinnate inflorescence, up to 80 cm tall, with reddish bracts and blue to violet corollas. *Tillandsia multicaulis* Steud. also has bright reddish bracts and blue corollas, but a smaller rosette and several sessile spikes, not exceeding the leaves. *Tillandsia punctulata* Schltdl. & Cham. is of tank-atmospheric intermediate habit; the inflorescence is about as long or slightly longer than the leaves and composed of few, densely digitate spikes with red bracts and dark violet petals with a white apex. *Tillandsia juncea* (Ruiz & Pav.) Poir. has filiform, fasciculate leaves; the inflorescence is rather small, composed of few, dense spikes; the bracts are reddish but less conspicuous than in the congeners studied; and the petals are violet. Although not growing on more exposed branches, *T. juncea* is clearly the most xeric of the bromeliads and the only species exhibiting Crassulacean acid metabolism (CAM). In contrast to the other species that are hermaphroditic, *Catopsis sessiliflora* (Ruiz & Pavón) Mez is dioecious; the leaves form small and narrow tanks; the inflorescence is pinnate or bi-pinnate with ca. 1 cm long creamish flowers and small and inconspicuous green bracts. *Tillandsia deppeana* is largely monocarpic, and the other species are polycarpic. The average number of flowers per ramet or individual and the diameter of branches, upon which reproductively mature individuals were growing are presented in Table 4.1.

Plants with developing inflorescences but without open flowers were collected in the forest with as much substrate (branch or bark) as possible and transferred ca. 50 m from the forest edge to the Botanical Garden Clavijero, where they were watered only by rainfall and were not fertilized. Light conditions were similar to those in the mid canopy, but more uniform than in the forest. Transplantation stress and differences in microclimate or pollinator presence may affect the reproductive output. This effect, however, was considered insignificant, since the pollination experiments in the garden were designed mainly to test for self-pollination and self-compatibility and not for reproductive success under natural conditions. Between March 2002 and May 2003, controlled pollination experiments were conducted for all species except *Tillandsia deppeana*, where inflorescences died before producing fruits and dioecious *Catopsis*.

Young flowers, spikes, or entire inflorescences received the following treatments: (A) some were emasculated, enclosed in fine-mesh bags, and pollinated with pollen from the same flower; (B) some were emasculated, bagged, and self-pollinated with pollen from a different flower of the same individual; (C) some were emasculated, bagged, and cross-pollinated with pollen from another plant; and (D) some were left untouched and bagged. With the very small flowers of *Jacquiniella*, hand-pollination was not possible; and flowers were either bagged and untouched (D) or remained open (E) as controls. Between August 2001 (wet season) and February 2004 (dry season), to assess pollination success under field conditions, counts were made of peduncles or dried flowers and fruits on orchid individuals that are part of a long-term population study. Because, in this study, observations of bromeliad inflorescences were not made over an extended period, we counted the number of flowers per ramet once (Table 4.1) and the number of fruits on different ramets later in the year, and calculated fruit set as the average number of capsules/average number of flowers per ramet.

The phenology of flowering and fruiting was observed by qualitatively recording the state of inflorescences along trails in the field in bi-weekly intervals between April 2002 and March 2003.

To assess the impact of branch size and stability on survival, we compared the branch-size distribution of reproductively mature individuals (data from Buchberger 2004) with the probability that a branch of a given size breaks within a year (data from Hietz 1997) to calculate the probability for a reproducing individual of a species to fall with its supporting branch within a year (Table 4.1).

Table 4.1: Average number of flowers per flowering shoot (orchids) or ramet (bromeliads), average diameter of branches occupied by reproductively mature individuals, and probability of a reproductively mature individual to fall with its breaking branch within a year.

Epiphyte species	Flowers per shoot or ramet No. (SD; N)	Branch diameter cm (SD; N)	Probability of epiphyte falling with branch
Bromeliaceae			
Catopsis sessiliflora	29.3 (22.8; 41)	3.9 (4.7; 88)	0.205
Tillandsia deppeana	74.3 (11.7; 8)	5.5 (2.5; 20)	0.120
Tillandsia juncea	16.4 (7.2; 17)	11.1 (8.1; 635)	0.063
Tillandsia multicaulis	12.9 (5.6; 19)	7.8 (6.7; 606)	0.106
Tillandsia punctulata	13.4 (7.4; 34)	8.4 (4.9; 794)	0.079
Orchidaceae			
Jacquiniella leucomelana	2.6 (27; 188)	7.4 (4.9; 43)	0.115
Jacquiniella teretifolia	2.9 (3.2; 39)	7.6 (8.0; 432)	0.103
Lycaste aromatica	4.3 (3.1; 157)	16.8 (9.6; 22)	0.023

Source: Data on branch diameter (Buchberger 2004), data on branch stability (Hietz 1997).
Note: SD = Standard deviation; N = number of plant

4.3 Results

Flowering phenology was diverse. The deciduous *Lycaste aromatica* flowered at the beginning of the rainy season prior to leaf production, *Jacquiniella teretifolia* towards the end of the rainy season, *J. leucomelana* during most of the year, *Tillandsia punctulata* and *T. deppeana* mainly during the dry season, and *T. multicaulis* and *Catopsis* during the wet season (Figure 4.1). Fruits had matured and seeds were dispersing mostly from the dry season to the beginning of the rainy season.

Out of the three species of *Tillandsia* tested, one (*T. juncea*) is self-compatible, with fruit sets similar in self- and cross-pollinated flowers (fruit sets 67–89%, Table 4.2). *Tillandsia juncea* also is capable of self-pollination, though self-pollination resulted in a somewhat lower fruit set (48%). *Tillandsia punctulata* and *T. multicaulis*, which are clearly not self-pollinating, also are largely self-incompatible, with self-pollination resulting in only about 3% of flowers producing fruits. Whether flowers were pollinated with pollen from the same flower, or from a different flower but the same individual (treatments A and B) made no difference. Whereas fruit sets of bagged *Jacquiniella teretifolia* and open controls were equally high, only 2 out of 57 flowers in bagged *J. leucomelana*

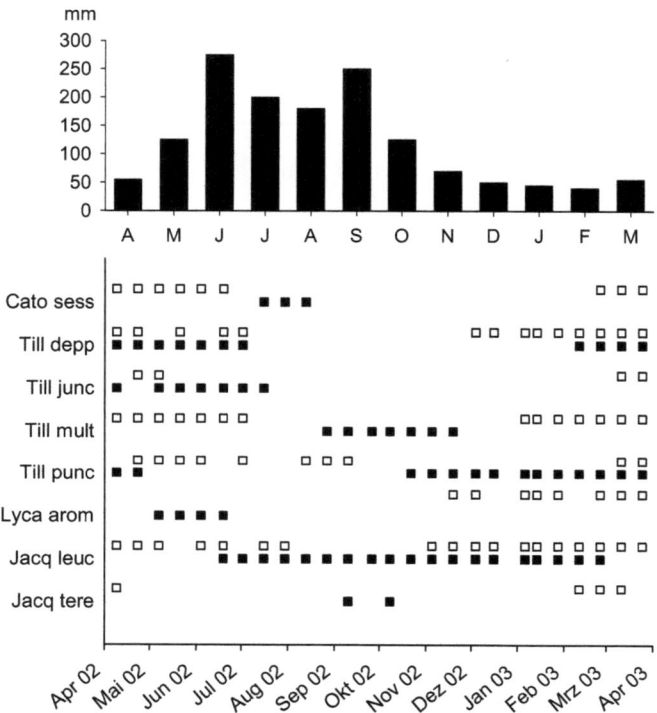

Figure 4.1: Rainfall (top) and phenology (bottom) of flowering (full symbols) and presence of mature seeds (empty symbols) of epiphytes studied in a Mexican humid montane forest. For species abbreviations see Table 4.1.

self-pollinated, but pollination of open controls was also low in this species. In *Lycaste aromatica*, no self-pollination occurred, and only one flower each in treatments A and B set fruit, whereas 39% of cross-fertilizations were successful (Table 4.2).

Selfing also can result in lower fruit set in species that are not 100% self-incompatible. This effect is quantified by the index of self-incompatibility, which is the ratio of the fruit set of hand self-pollinated to the fruit set of hand cross-pollinated flowers (Bullock 1985). By combining self-pollination and geitonogamy (treatments A and B), the index of self-incompatibility is >1 for *Tillandsia juncea*, 0.06 for *T. multicaulis*, 0.09 for *T. punctulata* and 0.11 for *Lycaste aromatica*. Natural fruit set (Table 4.3) in the potentially self-pollinating *Tillandsia juncea* was somewhat higher than with experimental exclusion of pollinators. Fruit set in the other species that are not selfing was between 70% in dioecious *Catopsis* and 25% in *T. punctulata*. Natural fruit set among the orchids was always lowest in *Lycaste aromatica* (8–11%); in *Jacquiniella teretifolia*, it was slightly higher than in the bagged and open controls exposed in the Botanical Garden (76–88%); and in *J. leucomelana*, it was substantially higher (29–40%) than in the experimental plants (Figure 4.2). In the orchids, fruit set was relatively constant in three consecutive years, though *J. leucomelana* could not be evaluated during the last observation in the dry season when flowers of this species were largely absent.

Table 4.2: Results of pollination experiments with the following treatments: **A.** Emasculated, bagged, and self-pollinated with pollen of the same flower. **B.** Emasculated, bagged, and self-pollinated with pollen from a different flower of the same individual. **C.** Emasculated, bagged, and cross-pollinated with pollen from another plant. **D.** Untouched and bagged. **E.** Open controls, exposed in the Botanical Garden Clavijero, Veracruz, Mexico.

Epiphyte species	Treatment	No. flowers / individuals	Fruit set Mean	(SD)
	A	28 / 5	0.67	(0.41)
	B	35 / 5	0.89	(0.15)
	C	37 / 13	0.67	(0.38)
Tillandsia juncea	D	167 / 6	0.48	(0.25)
	A	45 / 15	0.03	(0.13)
	B	50 / 13	0.02	(0.06)
	C	42 / 11	0.56	(0.40)
Tillandsia multicaulis	D	94 / 5	0.00	(0.00)
	A	42 / 10	0.00	(0.00)
	B	41 / 14	0.06	(0.18)
	C	42 / 15	0.42	(0.44)
Tillandsia punctulata	D	64 / 5	0.00	(0.00)
	D	57 / 11	0.02	(0.04)
Jacquiniella leucomelana	E	171 / 27	0.16	(0.28)
	D	37 / 25	0.75	(0.43)
Jacquiniella teretifolia	E	34 / 16	0.72	(0.37)
	A	24 / 6	0.08	(0.20)
	B	21 / 3	0.07	(0.12)
	C	25 / 4	0.39	(0.28)
Lycaste aromatica	D	24 / 4	0.00	(0.00)

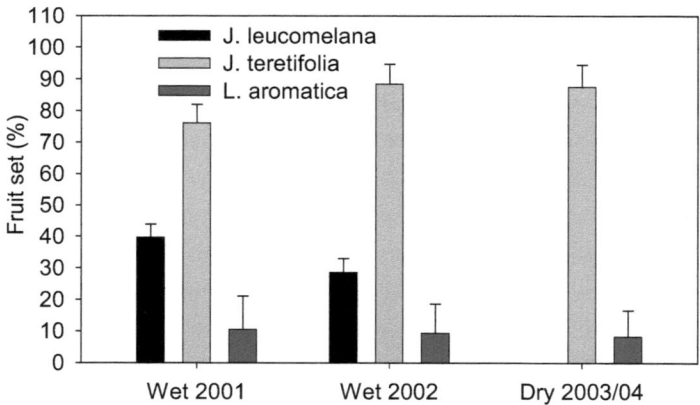

Figure 4.2: Average (± SE) fruit set of *Lycaste aromatica*, *Jacquiniella leucomelana*, and *J. teretifolia* in a natural forest during a 3-year period.

Table 4.3: Numbers of flowers and capsules per inflorescence and fruit set under natural conditions in epiphytic bromeliads.

Epiphyte species	Flower buds / inflorescence No. (SD; N)	Capsules / inflorescence No. (SD; N)	Average fruit set
Catopsis sessiliflora	29.3 (22.8; 41)	20.4 (10.6; 62)	0.706
Tillandsia deppeana	66.1 (26.1; 9)	44.4 (40.2; 31)	0.597
Tillandsia juncea	16.4 (7.2; 17)	9.8 (9.4; 69)	0.599
Tillandsia multicaulis	12.9 (5.6; 19)	5.3 (4.4; 93)	0.412
Tillandsia punctulata	13.4 (7.4; 34)	3.4 (2.5; 33)	0.254

4.4 Discussion

A prolonged flowering season could increase the probability of pollination, which should be an advantage for species growing on ephemeral substrates that may not live to the next flowering season. Among the species on thinner branches *Jacquiniella leucomelana* flowered for most of the year, but flowers of *Catopsis* and *J. teretifolia* each were observed only during a 2-month period; and the length of the flowering season did not appear to be related to habitat preferences.

Large inflorescences or showy flowers may attract more pollinators, but no apparent relationship was found between flower or inflorescence size and fruit set. Flower or inflorescence size is relatively large in *Lycaste* and *Tillandsia deppeana*, small in *Catopsis* and *Jacquiniella* spp., and intermediate in the other bromeliads.

The eight epiphytes investigated have diverse breeding systems: *Catopsis sessiliflora* is dioecious, *Tillandsia multicaulis*, *T. punctulata*, and *Lycaste aromatica* are largely or entirely self-incompatible and outcrossing, and *T. juncea* and *Jacquiniella teretifolia* are partly or mainly self-pollinating. The low fruit set of bagged and the moderate fruit set of openly pollinated *J. leucomelana* would suggest this species to be outcrossing. Observations that the minute flowers are nearly closed, do not produce nectar, and that fruit sets in the field were much higher than in the experimental control plants, questions the results of the pollination experiment and points to self-pollination, as reported earlier for this species (Catling 1990). The mesh bags of the pollination experiment (D) may have prevented self-pollination, if this is facilitated by rain or wind. Rain-facilitated selfing was reported for *Liparis loeselii* (Catling 1980), though the flower of *J. leucomelana* has no similarity with that species. Also a previous report that found *T. deppeana* self-compatible but not selfing (García-Franco & Rico-Gray 1991) is questionable; those pollination experiments largely failed; and not only the self-pollination, but also the cross-pollination produced no fruit at all.

Natural fruit set in bromeliads was highest in *Catopsis*, followed by *Tillandsia juncea*, *T. deppeana*, *T. multicaulis*, and *T. punctulata*. In the outcrossing *T. punctulata* and *T. multicaulis* natural fruit set was 74% and 60%, respectively, of hand-cross pollinated individuals; and natural fruit set was ≥60% in the other bromeliads. This is substantially higher than the fruit set in outcrossing *Lycaste aromatica* and in many other orchids (Neiland & Wilcock 1998, Tremblay et al. 2005), suggesting that bromeliads are less pollinator-limited than orchids. In *T. juncea*, fruit set in the field (60%) was not much higher than fruit set of only self-pollinating flowers (48%); thus a substantial proportion of flowers appears to be selfing under field conditions. This is the only atmospheric and certainly the most xeric species tested, and a high frequency of selfing concurs with the trend suggested for extreme epiphytes (Benzing 1978, Gilmartin & Brown 1985). A trend for selfing in xeric bromeliads also is reflected in isoenzyme variation, which shows strong inbreeding in the extremely xeric *Tillandsia recurvata*, a species with very small flowers. In contrast, the semi-xeric *T. ionantha*, which has long violet flowers and showy red bracts, is mainly or entirely outcrossing

(Soltis *et al.* 1987). The floral morphology of *T. punctulata*, where the stigma comes into close contact with the anthers, led to the assumption of self-pollination (Gardner 1986), but this is clearly not the case.

Benzing (2000) states that selfing is most conspicuous in monocarpic *Tillandsia* subgen. *Tillandsia*, but cites only the example of *T. utriculata*. Unfortunately, we could not test the breeding system of monocarpic *T. deppeana*; and an earlier report is dubious. Ensuring high fruit set through selfing or at least self-compatibility would be an obvious advantage for monocarpic species that, like annuals, cannot save resources not used by unpollinated flowers for future vegetative or generative growth. This hypothesis, however, awaits more pollination studies in monocarpic and polycarpic species. Apart from the breeding system, the very attractive inflorescence of *T. deppeana* and its production of ca. twice as much nectar per flower as *T. multicaulis* (Ordano & Ornelas 2004) should favor pollination resulting in the high fruit set observed.

Tillandsia deppeana, *T. multicaulis*, and *T. punctulata* were observed being visited by hummingbirds. At least *T. deppeana* and *T. multicaulis* also are visited by bees and butterflies and respond to repeated nectar removal by producing >3 times the nectar than unvisited flowers (Ordano & Ornelas 2004). Their flowers, therefore, are likely to be visited more than once and by more than one pollinating species, resulting in a high fruit set. *Catopsis sessiliflora*, while not of xeric habit, tends to grow on smaller and less stable branches than do the other bromeliads studied, with a corresponding high mortality of reproductive individuals (TABLE 1) and selective pressure to ensure high fruit set. Though the fruit set of *Catopsis* was indeed high, no responsible mechanism was obvious. The genus *Catopsis* includes dioecious and hermaphroditic species. Also within the species *C. sessiliflora*, some populations have perfect flowers, and others are dioecious (Benzing 2000). The population studied was entirely dioecious and therefore obligate outcrossing in contrast to xeric species of *Tillandsia* or *Jacquiniella* growing on relatively unstable branches. Although we did not observe or find reports on pollinators, *Catopsis sessiliflora*, which is certainly pollinated by insects, has the least conspicuous floral display of the bromeliads studied. This contrasts particularly with the low success of *Lycaste aromatica*, which has relatively large (ca. 5 cm), strongly scented, and dark yellow flowers, lasting for many days and pollinated by euglossine bees. While the density of *Lycaste* in the forest is not high, plants tend to grow in clusters on individual branches or trees, so that finding another flowering plant should not be difficult for the bees. Several species of orchids pollinated by Euglossinae and Melpomini bees have low fruit set in spite of frequent visitation, which is explained by incompatibility, geitanogamous pollination by bees remaining for several minutes at individual inflorescences and low genetic variability (Singer 2001, Singer & Koehler 2003). If neighboring individuals of *Lycaste* are closely related and share the same alleles determining incompatibility, many cross-pollinations may not result in fertile fruits. This, together with reliance on a specific pollinator, can explain the low fruit set.

Whether resulting from rare pollinators or self-incompatibility, fruit set may be less limiting for *Lycaste*, which is a long-lived species growing on large and rather stable branches, than it is for *Jacquiniella* spp. and *Catopsis*, growing on less stable substrate. Two other orchids, studied in a nearby coffee plantation and also fairly common in the forest, confirm this pattern. While *Maxillaria densa*, which prefers thicker branches, was largely self-incompatible and had a natural fruit set of 18.2%, *Scaphyglottis livida*, growing on thinner and more exposed branches, was self-compatible, though not autogamous, and had a natural fruit set of 35.5 % (Solis-Montero *et al.* 2005).

The time-limitation hypothesis, which states that selfing has evolved as a consequence of strong selection in ephemeral habitats, was confirmed in a review of annual plants, where selfing is generally widespread and occurs in higher frequencies in two of the most severely time-limited habitats—deserts and cultivated habitats (Snell & Aarssen 2005). In epiphytes, in addition to the temporal constraint on ephemeral branches, the potential investment in pollinator attraction is limited by resource availability, which is particularly true for xeric species adapted to dry forests or microhabitats.

In general, results from species studied here and by others (Gilmartin & Brown 1985, Solis-Montero *et al.* 2005) suggest that perennial epiphytes preferring the more ephemeral branches tend to ensure pollination by being autogamous, or at least increase the chances of pollination by being self-compatible. Breeding systems as well as habitat preferences, however, may also be related to phylogenetic groups. As an example, twig epiphytes among the orchids are mostly related genera within the Oncidiinae and appear to have smaller genome sizes, which may be an advantage in ephemeral and stressful habitats (Chase *et al.* 2005). Also, self-incompatibility and self-pollination are unevenly distributed within the orchids (Catling 1990, Borba *et al.* 2001, Tremblay *et al.* 2005). To sort out which traits are primarily phylogenetic, possibly representing pre-adaptations enabling the colonization of certain habitats, and which have evolved in response to specific habitats, data from a larger number of species is needed.

Epiphyte pollination has received early and widespread attention (Darwin 1888, van der Pijl & Dodson 1966) and remains a worthwhile subject for the study of breeding systems and pollination strategies as adaptations to specific environmental constraints. The present study, while generally supporting the time-limitation hypothesis, does leave open the intriguing question—what is the function of showy flowers, pollinator rewards, and breeding systems that promote pollination, if *Catopsis sessiliflora* achieved the highest fruit set in our study with guaranteed outcrossing and little apparent investment in reproductive structures or pollinator attraction?

4.5 Acknowledgements

We are grateful to the staff of the Botanical Garden Francisco Clavijero of the Instituto de Ecología in Xalapa and to José García-Franco for general support and two anonymous reviewers for helpful comments. This research was funded by the Austrian Science Fund (FWF grant number P14775 and P17875).

4.6 Literature

Aarssen L. W. 2000. Why are most selfers annuals? A new hypothesis for the fitness benefit of selfing. *Oikos* **89**: 606–612.

Ackerman J. D. 1986. Coping with the epiphytic existance. Pollination strategies. *Selbyana* **9**: 52-60.

Ackerman J. D. & Montalvo A. M. 1990. Short- and long-term limitations to fruit production in a tropical orchid. *Ecology* **71**: 263-272.

Ashman T.-L., Knight T. M., Steets J.A., Amarasekare P., Burd M., Campbell D. R., Dudash M. R., Johnston M. O., Mazer S. J., Mitchell R. J., Morgan M. T. & Wilson W. G. 2004. Pollen limitation of plant reproduction: ecological and evolutionary causes and consequences. *Ecology* **85**: 2408–2421.

Bartareau T. 1995. Pollination limitation, costs of capsule production and the capsule-to-flower ratio in *Dendrobium monophyllum* F. Muell, (Orchidaceae). *Australian Journal of Ecology* **20**: 257-265.

Bawa K. S. 1974. Breeding systems of tree species of a lowland tropical rain forest. *Annual Review of Ecology and Systematics* **21**: 399-422.

Benzing D. H. 1978. The life history of *Tillandsia circinnata* (Bromeliaceae) and the rarity of extreme epiphytism among the angiosperms. *Selbyana* **2**: 325-327.

Benzing D. H. 1990. *Vascular epiphytes – general biology and related biota*. Cambridge: Cambridge University Press.

Benzing DH. 2000. *Bromeliaceae: Profile of an Adaptive Radiation*. Cambridge: Cambridge University Press.

Borba E. L., Semir J. & Shepherd G. J. 2001. Self-incompatibility, inbreeding depression, and crossing potential in five Brazilian Pleurothallis (Orchidaceae) species. *Annals of Botany* **88**: 89–99.

Buchberger G. 2004. *Dreidimensionale Verteilung von epiphytischen Bromelian und Orchideen in einem humiden Bergwald Mexikos*. Diploma Thesis. Universität für Bodenkultur, Vienna.

Bullock S. H. 1985. Breeding systems in the flora of a tropical deciduous forest in Mexico. *Biotropica* **17**: 287-301.

Bush S. P. & Beach J. H. 1995. Breeding systems of epiphytes in a tropical montane wet forest. *Selbyana* **16**: 155-158.

Calvo R. N. 1993. Evolutionary demography of orchids: Intensity and frequency of pollination and the cost of fruiting. *Ecology* **74**: 1033-1042.

Calvo R. N. & Horvitz C. C. 1990. Pollinator limitation, cost of reproduction, and fitness in plants: A transition-matrix demographic approach. *American Naturalist* **136**: 499-516.

Catling P. M. 1980. Rain assisted autogamy in *Liparis loeselii* (L.) L.C.Rich. (Orchidaceae). *Bulletin Torrey Bot. Club* **4**: 525–529.

Catling P. M. 1990. Auto-pollination in the Orchidaceae. Pp. 121-158 in: Arditti J. (ed.). *Orchid Biology, Reviews and Perspectives V*. Portland, Oregon:Timber Press.

Chase M. W., Hanson L., Albert V. A., Whitten W. M. & Williams N. H. 2005. Life history, evolution, and genome size in subtribe Oncidiinae (Orchidaceae). *Annals of Botany* **95**: 191–199.

Darwin C. 1888. *The Various Contrivances by Which Orchids Are Fertilised by Insects*. 2nd. edition, John Murray, London.

Dressler R. L. 1968. Pollination by euglossine bees. *Evolution* **22**: 202-210.

Dressler R. L. 1981. *The orchids: natural history and classification*. Cambridge: Harvard University Press.

García-Franco J. G. & Rico-Gray V. 1991. Reproductive biology of *Tillandsia deppeana* (Bromeliaceae) in Veracruz, Mexico. *Brenesia* **35**:61-79.

Gardner C. S. 1986. Inferences about pollination in *Tillandsia* (Bromeliaceae). *Selbyana* **9**: 76-87.

Gilmartin J. A. & Brown G. K. 1985. Cleistogamy in *Tillandsia capillaris* (Bromeliaceae). *Biotropica* **17**: 256-242.

Hietz P. 1997. Population dynamics of epiphytes in a Mexican humid montane forest. *Journal of Ecology* **85**: 767-775.

Hietz P., Ausserer J. & Schindler G.. 2001. Growth, maturation and survival of epiphytic bromeliads in a humid montane forest. *Journal of Tropical Ecology* **18**: 177–191.

Hietz P. & Hietz-Seifert U. 1995. Intra- and interspecific relations within an epiphyte community in a Mexican humid montane forest. *Selbyana* **16**: 135-140.

Holdridge L. R. 1967. *Life zone ecology*. San José, Costa Rica: Tropical Science Center.

Janzen D. H. 1977. A note on optimal mate selection in plants. *American Naturalist* **111**: 365–371.

Johnsen K., Major J. E. & Maier C. A. 2003. Selfing results in inbreeding depression of growth but not of gas exchange of surviving adult black spruce trees. *Tree Physiology* **23**: 1005–1008.

Kessler M. & Krömer T. 2000. Patterns and ecological correlates of pollination modes among bromeliad communities of Andean forests in Bolivia. *Plant Biology* **2**: 659-669.

Kress W. J. & Beach J. H. 1994. Flowering plant reproductive systems. In McDade L. A., Bawa K. S., Hespenheide H. A. & Hartshorn G. S. (eds.) *La Selva. Ecology and Natural History of a Neotropical Rain Forest*. Pp. 161–182. London: University of Chicago Press.

Lumer C. 1980. Rodent pollination of Blakea (Melastomataceae) in a Costa Rican cloud forest. *Brittonia* **32**: 512-517.

Martinelli G. 1994. *Reproductive Biology of Bromeliaceae in the Atlantic Rainforest of Southeastern Brazil*. University of St. Andrews. St. Andrews, Scotland.

Melendez-Ackerman E. J. , Ackerman J. D. & Rodriguez-Robles J. A. 2000. Reproduction in an orchid can be resource-limited over its lifetime. *Biotropica* **32**: 282-290.

Montalvo A. M. & Ackerman J. D. 1987. Limitations to fruit production in *Ionopsis utricularioides* (Orchidaceae). *Biotropica* **19**: 24-31.

Neiland M. R. & Wilcock C. C. 1998. Fruit set, nectar reward, and rarity in the Orchidaceae. *American Journal of Botany* **85**: 1657-1671.

Ordano O. M. & Ornelas J. F. 2004. Generous-like flowers: nectar production in two epiphytic bromeliads and a meta-analysis of removal effects. *Oecologia* **140**: 495-505.

Rzedowski J. 1986. *Vegetación de México*. (Third edition). México: Editorial Limusa. 432 pp.

Singer R. B. 2001. The pollination of *Trigonidium obtusum* Lindl. (Orchidaceae: Maxillariinae): trap-flowers and sexual mimicry. *Annals of Botany* **89**: 157–163.

Singer R. B. & Koehler S. 2003. Notes on the pollination biology of *Notylia nemorosa* (Orchidaceae): do pollinators necessarily promote cross pollination? *Journal of Plant Research* **116**:25.

Snell R. & Aarssen L. 2005. Life history traits in selfing versus outcrossing annuals: exploring the 'time-limitation' hypothesis for the fitness benefit of self-pollination. BMC *Ecology* **5**: 2.

Solis-Montero L., Flores-Palacios A. & Cruz-Angón A. 2005. Shade-coffee plantations as refuges for tropical wild orchids in central Veracruz, Mexico. *Conservation Biology* **19**: 908–916.

Soltis D. E., Gilmartin A. J., Rieseberg L. & Gardner S. 1987. Genetic variation in the epiphytes *Tillandsia ionantha* and *T. recurvata* (Bromeliaceae). *American Journal of Botany* **74**: 531-537.

Tanaka Y. 1997. Extinction of populations due to inbreeding depression with demographic disturbances. *Researches on Population Ecology* **39**: 57–66.

Tremblay R. L., Ackerman J. D., Zimmerman J. K. & Calvo R. N. 2005. Variation in sexual reproduction in orchids and its evolutionary consequences: a spasmodic journey to diversification. *Biological Journal of the Linnean Society* **84**: 1–54.

van der Pijl L. & Dodson C. H. 1966. *Orchid Flowers: Their Pollination and Evolution*. Coral Gables, Fla.: University of Miami Press.

Ward M., Dick C. W., Gribel R. & Lowe A. J. 2005. To self, or not to self... A review of outcrossing and pollen-mediated gene flow in neotropical trees. *Heredity* **95**: 246–254.

Williams-Linera G. 1997. Phenology of deciduous and broadleaved-evergreen tree species in a Mexican tropical lower montane forest. *Global Ecology and Biogeography Letters* **6**: 115-127.

Zimmerman J. K. & Aide T. M. 1989. Patterns of fruit production in a Neotropical orchid: Pollinator vs. resource limitation. *American Journal of Botany* **76**: 67-73.

Zotz G. 1998. Demography of the epiphytic orchid, *Dimerandra emarginata*. *Journal of Tropical Ecology* **14**: 725–741.

Zotz G. & Schmidt G. 2006. Population decline in the epiphytic orchid *Aspasia principissa*. *Biological Conservation* **129**: 82–90.

5 Spatial and temporal variability in the population dynamics of epiphytic orchids in a Mexican humid montane forest

Manuela Winkler, Karl Hülber and Peter Hietz

Abstract

Epiphytes need to balance the chances of long-term survival and those of colonizing new branches in their high-stress and high-risk canopy habitats.

We censused populations of three orchid species over three years, and calculated complete matrix population models to determine the key factors influencing population growth rates, and logistic regression models to study how demographic features change with microsite characteristics.

Average asymptotic population growth rates (λ_1) were below unity in all species except for *Jacquiniella teretifolia* in one of the years studied. The juvenile stages dominated the stable-stage distributions and the actual frequencies in most species. The time from seed germination to a fertile individual ranged between six and 16 years. Reproductive values increased with plant size.

Population growth rates were most sensitive to the survival of large pre-reproductive or reproductive stages. Fecundity and seedling survival had only a small effect.

Mortality decreased with age with the exception of *Jacquiniella teretifolia* where many reproductive plants died when one branch fell. The probabilities of surviving and of growing from the pre-reproductive to the reproductive stage increased with the amount of light available in most species.

Most of the populations studied are declining. In combination with the ever decreasing amount of natural habitats this may lead to the extinction of some of the species, especially those restricted to closed forests.

5.1 Introduction

A significant part of the science of ecology is concerned with trying to understand what determines the abundance of organisms. To provide answers to this question various demographic methods are applied to population census data (Begon, Harper & Townsend 1990). Among these, matrix population models are well established to study aspects of species conservation (Bergman et al. 1993, Hodgson & Townley 2004, Wisdom et al. 2000), spatiotemporal variation, and metapopulation biology (Guardia et al. 2000, Valverde & Silvertown 1997, 1998), and the influence of environmental stochasticity on populations (Aberg 1992, Hoffmann 1999, Pascarella & Horvitz 1998, for an overview see Caswell 2001, and the references therein).

Vascular epiphytes are a conspicuous and characteristic element of tropical moist forests. They account for a substantial proportion of tropical plant diversity and are important factors in ecosystem processes (Benzing 1990). However, in contrast to tropical trees and herbs (e. g. Martinez-Ramos & Sarukhán 1984, Horvitz & Schemske 1995, Olmsted & Alvarez-Buylla 1995), very few complete population models for epiphytes have been presented. Some aspects of epiphyte population biology have been studied by Benzing (1978, 1981), Bennett (1988, 1991), Larson (1992), and Zotz (1998). Hietz (1997) presented a study on epiphyte population dynamics based on repeated photographs of branches. Calvo (1993) investigated the trade-off between current reproduction and future growth or reproduction in an epiphytic orchid using a matrix-model. Effective population sizes of populations of epiphytic orchids of the genus Lepanthes were estimated using a matrix-model approach (Tremblay & Ackerman 2001). Hernández-Apolinar (1992) and Tremblay (1997) used Lefkovitch matrices to describe populations of the epiphytic orchids *Laelia speciosa* and *Lepanthes caritensis*. Mondragón et al. (2004) provided the only complete matrix population model for an epiphyte, the bromeliad *Tillandsia brachycaulos*, covering spatial and temporal variability.

In contrast to ground-rooted plants, the population dynamics of epiphytes is complicated by the fact that the canopy is a very heterogeneous and fragmented habitat with high rates of patch turnover (Benzing 1990, Hietz 1997). On the one hand, forest canopies are high-risk habitats, which should favour investment in high reproduction, on the other hand reproductive power is reduced in many epiphytes that need to invest in stress tolerance (Benzing 1990). Epiphytes thus need to balance the long-term retention of existing anchorage sites against the recruitment of new ones.

In this study we present complete matrix population models for three species of epiphytic orchids. We addressed the following questions: (1) What is the demographic pattern in these species? (2) Which are the key demographic processes determining population growth rates? (3) How are populations affected by canopy microsites and how do they change through time?

5.2 Material and Methods

Study area and species

This study was conducted in the forest reserve 'Santuario Bosque de Niebla' adjacent to the Instituto de Ecología, 2.5 km south of Xalapa, in central Veracruz, Mexico (19°31'N, 96°57'W), at 1350 m elevation. Average temperature is 19° C, and annual precipitation is 1500 mm, most of which falls in the June to October rainy season. The forest is at the transition between premontane and lower montane moist forest according to the Holdridge (1967) life-zone system. In Mexico, it is commonly classified as 'bosque mesófilo de montaña' (mesophilous montane forest) following Rzedowski (1986). Descriptions of the forest structure are given by Williams-Linera (1997) and of the epiphyte community by Hietz and Hietz-Seifert (1995).

We censused populations of the epiphytic orchids *Jacquiniella leucomelana* (Reichenbach f.) Schlechter, *J. teretifolia* (Sw.) Britton & P. Wilson and *Lycaste aromatica* (Graham ex Hook.) Lindley.

Population census

Orchid populations were tagged on ten trees with a maximum distance between two trees of ca. 250 m in August 1999 and 2001. To account for their clumped distribution (Hietz & Hietz-Seifert 1995, Buchberger 2004), all individuals on a tree were included, with the exception of those on inaccessible branches. In the *Jacquiniella* species, the number of ramets, flowers, dehisced flowers and fruits was recorded, and the length of the longest ramet was measured to the nearest 0.5 cm. In *Lycaste aromatica*, the number of peduncles and fruits was recorded. The height, breadth and width of each living pseudobulb were measured in the first census, and the height of the most recent pseudobulb in the subsequent censuses. Assuming that an ellipsoid fits the shape of a pseudobulb best, pseudobulb volume was calculated as

$H/2 * B/2 * W/2 * 4 * \pi / 3$

where H, B and W are pseudobulb height, breadth and width, respectively. Pseudobulb width and length are linear correlated with height ($W = H*0.62$ and $L = H*0.35$, compare Winkler 2001).

Branch height relative to total tree height and distance to the trunk relative to the crown radius, inclination, circumference, and canopy openness (using a convex spherical densiometer, Ben Meadows, Atlanta, GA), were measured and percent bryophyte and total epiphyte plant cover were estimated for each branch. Survival, vegetative growth and reproduction of the tagged individuals were monitored and newly emerging seedlings were recorded. Orchid population were censused in February and August 2002, in February 2003, 2004 and 2005.

Matrix population models

To characterize population dynamics we used Lefkovitch (1965) stage-classified transition matrices. A combination of size (length of longest leaf) and reproductive criteria was used to distinguish stage classes (as in Menges 1990 and Horvitz & Schemske 1995; Table 5.1). The size threshold of seedlings was defined as the 95th percentile size of one year-old seedlings from census seedlings in the orchids. Pre-reproductives were defined as being at least of the size of the 5th percentile of reproductive plants, *i.e.* plants that had reached the size of reproducing individuals, but had not yet produced flowers. Three different reproductive stages were distinguished, based on the 33th and 67th percentile of the size of reproductive plants. Fecundity was calculated as the number of newly emerged seedlings at the end of a projection interval divided by the number of reproductive individuals at the beginning of the interval. Since seedlings could not be assigned to an individual mother-plant, orchid seedlings were distributed among reproductive classes according to the number of fruits produced by each fertile stage at the beginning of the projection interval (Figure 5.1, Table 5.1).

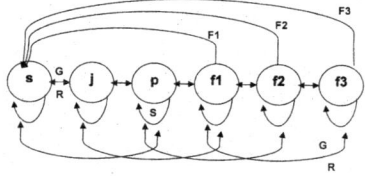 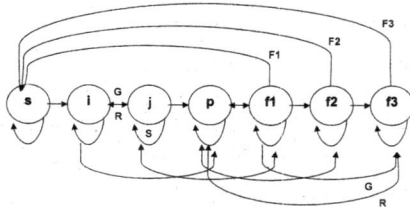

Figure 5.1: Life cycle graphs of the epiphytic orchids *Jacquiniella* species (left), and *Lycaste aromatica* (right). Circles indicate stage classes (see Table 5.1). F = Fecundity, S = Stasis, G = Growth, R = Retrogression.

Three 6 x 6 transition matrices were built for *Jacquiniella* spp., and three 7 x 7 matrices for *L. aromatica*, for the three growth periods studied. Additionally, all observations of particular transition events were pooled over years (Feb. 2002 – 2005) to create a pooled matrix for each species (compare Horvitz & Schemske 1995). The finite population growth rate λ_1 (the dominant eigenvalue of the transition matrix), the stage-specific reproductive value v (the left eigenvector, scaled so that first entry of v is 1) and the stable-stage distribution w (the right eigenvector, scaled so that the vector's sum is 1) were calculated using the *eigen.Matrix* command from the S-Plus Matrix library (S-plus 2000 MathSoft Inc. 1988-1999). Empirical confidence intervals for λ_1 were obtained using nonparametric bootstrapping (*bootstrap* command in S-Plus 2000) with n = 1000 resamples drawn (Efron & Tibshirani 1993, Caswell 2001).

The rate of convergence to the stable stage distribution is governed by the eigenvalue with the second largest magnitude (λ_2). Convergence will be more rapid the larger the dominant eigenvalue is relative to the other eigenvalues. The damping ratio p

$$p = \lambda_1 / |\lambda_2|$$

and t_x, the time required for the contribution of λ_1 to become x times that of λ_2

$$t_x = \log(x)/\log(p)$$

were calculated as measures of the rate of convergence (Caswell 2001).

Keyfitz's distance measures the distance between two vectors (*i.e.* between the stable-stage and the observed stage distribution). It is calculated as

$$\Delta(x,w) = \frac{1}{2} \sum_i |x_i - w_i|$$ where x is the observed stage distribution.

It varies between 0 (vectors are identical, *i.e.* the observed stage is stable) and 1 (vectors are completely different) (Caswell 2001).

The matrix N gives the mean time spent in each stage. It is calculated as the inverse of the identy matrix minus the transition matrix (for details see Caswell 2001, Chapter 5). Adding the entries on the diagonal until reaching the first reproductive stage gives the time to first reproduction (t_{rep}), and the column sums give the mean time to death for each stage (t_d).

Elasticity matrices were constructed to evaluate the relative importance of each vital rate (matrix entry a_{ij}) to population growth rate. The elasticity of λ_1 is defined as

$$e_{ij} = \frac{a_{ij}}{\lambda_1} \frac{\delta \lambda}{\delta a_{ij}} = \frac{1}{\lambda_1} s_{ij} a_{ij}$$

where s_{ij} is the absolute sensitivity of λ_1 to changes in a_{ij} (Caswell 2001, de Kroon et al. 1986). The sum of the elasticity matrix entries equals one. The elasticity e_{ij} can be interpreted as the slope of log λ_1 plotted against log a_{ij} and measures proportional sensitivity (Caswell 2001). Elasticity values were added up across demographic processes (stasis – survival without changing stage, retrogression to smaller stages, growth and fecundity; Silvertown et al. 1993) and across life-cycle stages (Horvitz & Schemske 1995).

Logistic Regression Models

To evaluate the influence of branch properties on demographic traits we used simple logistic regression models to analyse the probability of growth from the pre-reproductive to the reproductive stage. For stage specific-mortality and influence of single branch parameters proportional odds models were applied. These are cumulative logit models simultaneously using all response categories in a single model. Each cumulative logit has its own intercept (α_j) increasing with j, but coefficients are the same (β):

$$\text{logit}\,[P(y \leq j \mid x)] = \alpha_j + \beta' x$$

where $j = 1, \ldots, J\text{-}1$ is the respective stage out of J stages, α_j are the intercepts, β is the vector of coefficients and x is the vector of branch parameters analysed (*lrm* command of the S-Plus Design-library, Agresti 2002, Anonymous 1999).

Table 5.1: Criteria to distinguish stage classes. Sizes are length of longest ramet in *Jacquiniella* spp. (cm) and volume of all living pseudobulbs in *Lycaste aromatica* (cm^3). Reproductive stages have flowers and/or fruits. Note that stages are defined species-specific (n.d. = not defined).

Species	1-s Seedling	2-i Infant	3-j Juvenile	4-p Pre-reproductive	5-f Reproductive	5-f1 Small Reproductive	6-f2 Large Reproductive	7-f3 Extra-large Reproductive
Jacquiniella leucomelana	<=1.5	n.d.	> 1.5 and <=3.0	> 3.0	n.d.	<=5.5	> 5.5 and <=7.5	>7.5
J. teretifolia	<=1.5	n.d.	> 1.5 and <=10.5	> 10.5	n.d.	<=20.5	> 20.5 and <=26.5	> 26.5
Lycaste aromatica	<=0.4	> 0.4 and <=16.0	> 16.0 and <=31.5	> 31.5	n.d.	<=68.0	> 68.0 and <=120.5	> 120.5

5.3 Results

Demographic properties

The life cycles of the study species display transitions that include stasis, growth by one or several stages, retrogression by one or two stages and sexual reproduction (Figure 5.1, Table 5.2). Reproductive values increased with plant size in all species for the pooled transition matrix and in most single-year matrices. In the projection interval 2002-2003, the reproductive value of *J. teretifolia* reached its maximum in the pre-reproductive stage (Appendix). Time to reproduction obtained from the pooled transition matrices ranged from ca. 6 years in *J. leucomelana* to 11-12 years in *J. teretifolia* and *L. aromatica* (Table 5.3).

The small stages dominated the stable-stage distributions of the *Jacquiniella* species. By contrast, in *L. aromatica* the largest reproductive stage accounted for more than 60 % of the population in the pooled matrix (Table 5.2). The actual stage distribution was dominated by the infant stage in all species except *Jacquiniella leucomelana* with a clearly dominant seedling stage (Table 5.2). According to Keyfitz's distance measure, the damping ratios and t_{10}-values, the current stage distribution of the orchids is close to the stable-stage distribution (Table 5.3).

Orchid seedling mortality was usually below or near 50 %, with the exception of *L. aromatica* in 2003-2004 when more than three quarters of the seedlings died (Table 5.2, Appendix). Mortality generally decreased with age, with the exception of *Jacquiniella teretifolia*. Mean time to death correspondingly increased with age in most species (Table 5.2). Variability among years was high in *Lycaste aromatica* and *Jacquiniella teretifolia* (Table 5.3).

The finite population growth rates of the pooled transition matrices were below unity in all species even when considering the confidence intervals. None of the species reached a population growth rate that was significantly higher than one, though *J. teretifolia* (in this case the confidence intervals could not be calculated) reached λ_1-values of 1.224 in the second year (Table 5.3). Variation among years was high in *Jacquiniella teretifolia*.

Table 5.2: Pooled transition matrices and demographic properties of populations. Stages as in Table 5.1, other symbols are q_x = mortality rate, v = stage-specific reproductive value, w = stable-stage distribution, t_d = stage-specific time to death (years). For detailed matrices of each year see Appendix.

pooled		s	i	j	p	f(1)	f2	f3	v	w	t_d	$\frac{n}{Feb.02}$
Jacquiniella leucomelana												
	s	0.361		0.076	0.009	0.381	0.548	0.943	1.000	0.349	3.13	321
	j	0.185		0.387	0.048	0.045	0.012	0.000	2.130	0.162	4.38	154
	p	0.037		0.237	0.338	0.311	0.245	0.231	2.899	0.223	4.44	141
	f1	0.003		0.078	0.155	0.299	0.074	0.007	3.650	0.094	4.92	62
	f2	0.001		0.002	0.111	0.147	0.331	0.044	4.736	0.078	5.25	67
	f3	0.001		0.000	0.072	0.011	0.202	0.532	5.256	0.095	4.89	136
	q_x	0.412		0.219	0.268	0.169	0.129	0.186				
J. teretifolia												
	s	0.400		0.014	0.000	0.364	0.715	1.823	1.000	0.181	5.91	33
	j	0.367		0.733	0.031	0.043	0.000	0.000	1.458	0.386	6.94	176
	p	0.000		0.094	0.766	0.565	0.341	0.139	2.930	0.353	8.04	91
	f1	0.000		0.002	0.031	0.087	0.024	0.000	2.642	0.015	6.69	17
	f2	0.000		0.000	0.048	0.043	0.268	0.069	3.701	0.030	6.15	24
	f3	0.000		0.000	0.027	0.000	0.098	0.583	7.059	0.035	6.10	34
	q_x	0.233		0.157	0.096	0.261	0.268	0.208				
Lycaste aromatica												
	s	0.234	0.003	0.000	0.000	0.022	0.041	0.083	1.000	0.073	4.80	26
	i	0.266	0.668	0.014	0.000	0.000	0.000	0.000	2.754	0.066	10.06	131
	j	0.000	0.129	0.254	0.000	0.012	0.000	0.000	5.970	0.012	17.00	30
	p	0.000	0.003	0.408	0.539	0.301	0.138	0.088	8.059	0.167	22.08	70
	f1	0.000	0.003	0.113	0.056	0.289	0.000	0.000	8.204	0.016	22.43	34
	f2	0.000	0.000	0.000	0.183	0.301	0.340	0.000	8.932	0.057	23.85	31
	f3	0.000	0.000	0.000	0.139	0.036	0.457	0.885	9.722	0.609	25.59	39
	q_x	0.500	0.194	0.211	0.083	0.060	0.064	0.027				

Table 5.3: Matrix properties: Population growth rate (λ_1 *observed*, and 5 % and 95 % confidence interval from bootstrap), damping ratio (ρ), time required for the contribution of λ_1 to become ten times as great as that of λ_2 (in years, t_{10}), Keyfitz's distance between the observed and the stable stage distribution ($Dist_{Key}$), time to reproduction (in years, t_{rep}). na = not available.

	λ_1 obs.	5 % CI	95 % CI	ρ	t_{10}	$Dist_{Key}$	t_{rep}
pooled							
Jacquiniella leucomelana	0.883	0.865	0.903	1.870	3.677	0.263	5.80
J. teretifolia	0.936	0.912	0.962	1.426	6.491	0.234	11.96
Lycaste aromatica	0.966	0.949	0.985	1.514	5.548	0.383	11.41
Feb. 2002 - Feb. 2003							
Jacquiniella leucomelana	0.884	0.846	0.920	1.733	4.188	0.229	6.41
J. teretifolia	0.852	na	na	1.173	14.450	0.389	11.61
Lycaste aromatica	0.976	0.947	1.000	1.596	4.926	0.282	14.38
Feb. 2003 - Feb. 2004							
Jacquiniella leucomelana	0.826	0.787	0.867	1.602	4.884	0.268	5.79
J. teretifolia	1.224	na	na	1.594	4.940	0.393	7.26
Lycaste aromatica	0.929	0.879	0.983	1.498	5.697	0.512	8.42
Feb. 2004 - Feb. 2005							
Jacquiniella leucomelana	0.912	0.879	0.949	1.867	3.687	0.228	4.81
J. teretifolia	0.887	0.851	0.938	1.137	17.993	0.335	13.94
Lycaste aromatica	0.997	0.980	1.013	1.537	5.361	0.393	17.07

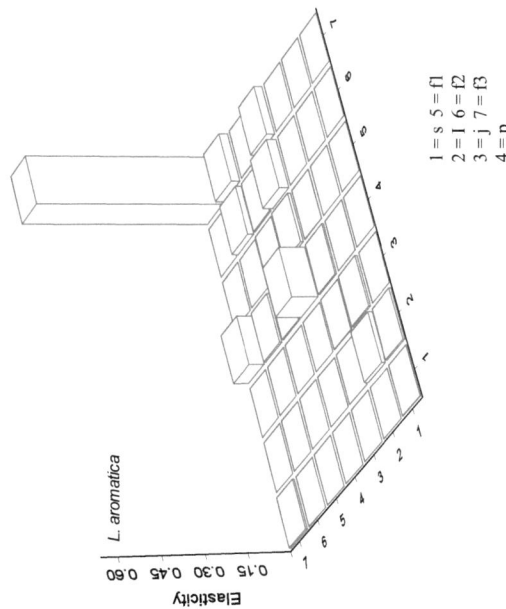

Figure 5.2: Elasticities of λ to changes in the entries p_{ij} of pooled transition matrices. Note the scale on the Z-axis. Stages as in Table 5.1.

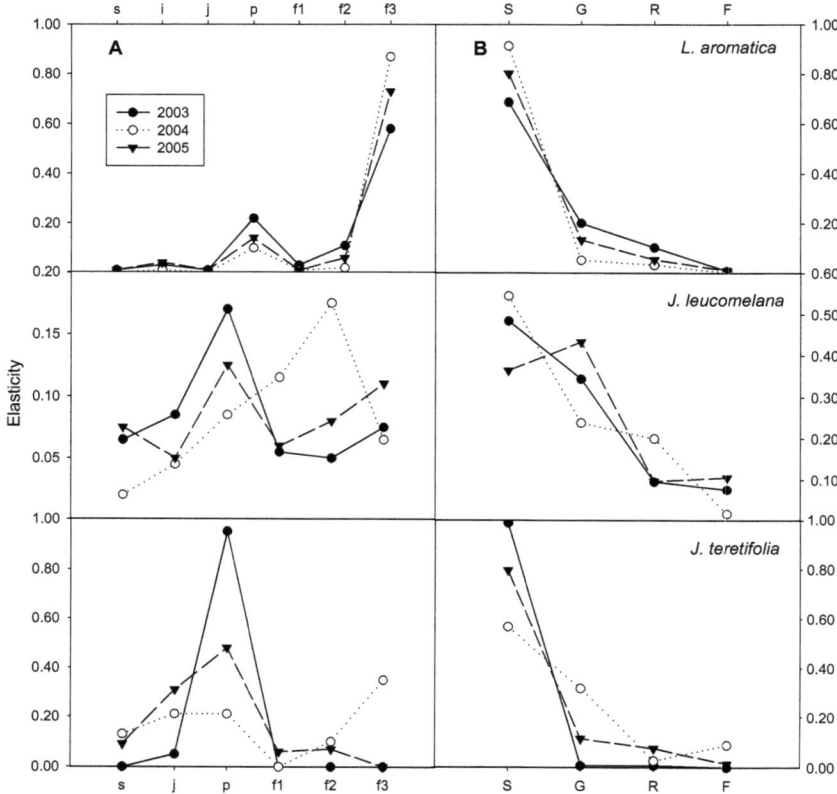

Figure 5.3: Added elasticity values A) by stage and B) by vital rate: S = stasis, G = growth, R = retrogression, F = fecundity for orchid transition matrices 2002-2005. Stages as in Table 5.1.

Elasticity analysis

Stasis was the most important demographic process in all species with stasis of reproductive and pre-reproductive stages as the vital rates with the highest elasticity values (Figure 5.2). The relative importance of fecundity, *i.e.* the transition from reproductive to seedling stage, was nearly zero in all species. The contribution of demographic processes varied very little among years with stasis being the most important vital rate followed by growth, retrogression and fecundity (Figure 5.3B). Extra-large reproductives (f3) had the greatest influence on λ_1 in *L. aromatica* in all three years analysed, followed by pre-reproductives with only one-fourth of the added elasticity values of the f3-stage (Figure 5.3A). For the population growth rate of the *Jacquiniella* species, the pre-reproductive stage was the most important with the exception of the projection interval 2003-2004 when large or extra-large reproductives had the greatest influence (Figure 5.3A).

Influence of branch characteristics

In *L. aromatica* 1.2 % of the transition events in the pooled matrices were deaths due to branchfall, 1.4 % in *J. leucomelana*, and 9.1 % in *J. teretifolia*, respectively. The influence of higher mortality because of branchfall on population growth rates was relatively low (Figure 5.4).

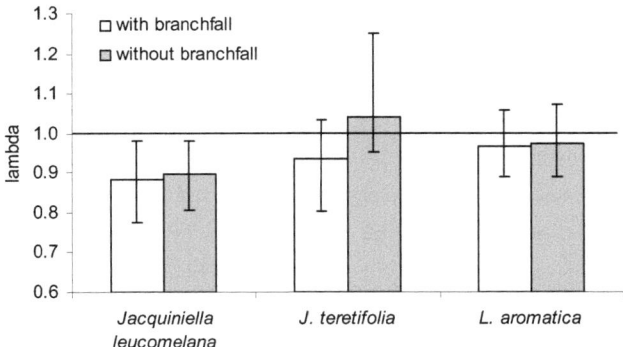

Figure 5.4: Population growth rate (λ) of pooled populations (observed, 5 % and 95 % empirical confidence interval from bootstrap) with and without individuals fallen with their supporting branch.

The probability that a given stage survived an observation interval was influenced by several characteristics of the supporting branch (proportional odds models, Table 5.4). Increasing canopy openness or relative height in the tree resulted in higher survival probabilities. Also, a higher bryophyte and plant cover increased survival probabilities in the dry season significantly, whereas for *Jacquiniella leucomelana*, high plant and bryophyte cover resulted in lower survival in both seasons (Table 5.4).

Similar to survival, the probability that a pre-reproductive plant becomes reproductive within a projection interval of one year increased with canopy openness and exposure (relative height and relative distance) (logistic regression models, Table 5.5). Increasing bryophyte cover resulted mostly in decreasing growth probabilities, above all for *Jacquiniella leucomelana*, as did branch size and inclination (Table 5.5).

5.4 Discussion

Demographic properties

The epiphytes of our study are slow-growing and reach fertility only after 6 – 12 years. *Lycaste aromatica* was estimated to become fertile at the age of ca. 11 years, which agrees well the ten years calculated by Winkler & Hietz (2001), who estimated the age of individuals through a simple calculation of the size difference on chains of the regularly produced pseudobulbs. Hietz *et al.* (2002), measuring size of bromeliads on repeated photographs of branches between 1992 and 1998, predicted five bromeliad species to reach fertility after 9 - 18 years.

The reproductive values of fertile orchids exceed those of seedlings by the factor 5 – 10. The reproductive values of the epiphytic orchid *Laelia speciosa* (Hernández-Apolinar 1992) were similar to those of *Lycaste aromatica*. For the epiphytic bromeliad *Tillandsia brachycaulos*, reproductive values of the largest sexually derived size class were only 3.6-fold higher than the seedlings value, and 4.8-fold for vegetatively derived fertile plants (Mondragón *et al.* 2004). Reproduction, survival, and timing all have an influence on reproductive value. Low values at birth reflect the probability that a seedling will die before reaching reproductive size and the delay before reaching it (Caswell 2001).

Table 5.4: Probability that fate is "alive" in the projection intervals February 2002 - August 2002 (dry season) and August 2002 - February 2003 (rainy season), respectively, depending on state and branch parameter. R² from proportional odds models are given for significant coefficients of branch parameters. Individuals fallen with their supporting branch were excluded from the analysis.

	Canopy openness		relative Height		relative Distance to trunk		Circumference		Inclination		Bryophyte cover		Plant cover	
	dry	rainy	dry	rainy	dry	rainy	dry	rainy	dry	rainy	dry	rainy	dry	rainy
Jacquiniella leucomelana	+	~	+	+	~	+	~	~	+	~	-	-	-	-
r²			0.096	0.214					0.106		0.091	0.177	0.092	0.19
J. teretifolia	+	+	-	-	-	-	+	+	~	+	+	~	-	~
r²	0.202	0.458			0.194	0.281	0.242	0.362		0.283		0.176		
Lycaste aromatica	+	~	+	+	~	+	-	-	~	~	+	+	-	+
r²								0.241			0.179		0.170	

+ = probability increasing with branch parameter, - = decreasing with parameter, ~ = branch parameter has no influence

Table 5.5: Probability that a pre-reproductive plant becomes reproductive within a projection interval (2002 - 2003; *J. teretifolia*: 2004 - 2005) depending on branch parameters. R² are given from logistic regression models with a significant coefficient for the single branch parameters. Only living individuals were included in the analysis.

	Canopy openness	relative Height	relative Distance	Circum-ference	Incli-nation	Bryo-phyte cover	Plant cover
Jacquiniella leucomelana	+	+	+	-	~	0.084	-
r²		0.066	0.062				0.096
J. teretifolia	+	+	~	~	~	-	-
Lycaste aromatica	+	+	-	~	~	-	-

+ = probability increasing with branch parameter, - = decreasing with parameter, ~ = branch parameter has no influence

The stable-stage distribution gives the proportion of individuals in each stage class to which the relative frequencies of stages will converge regardless of initial frequencies (Horvitz & Schemske 1995, Caswell 2001). The stable-stage distribution was skewed towards the seedling,infant or juvenile stage. The same was true for *Laelia speciosa* (Hernández-Apolinar 1992), and *Tillandsia brachycaulos* (Mondragón *et al.* 2004), where infant offshoots predominated. In *Calathea ovadensis*, a tropical understorey herb, seedlings made up more than 50 % of the non-seed stages in most years and plots (Horvitz & Schemske 1995), and 87.3 % and 56.3 % in two Mexican species of the genus *Neobuxbaumia* (Cactaceae; Godínez-Alvarez & Valiente-Banuet 2004). The seedlings of the temperate perennial herb *Primula vulgaris*, dominated the stable stage distribution only in large gaps (Valverde & Silvertown 1998). In the epiphytic orchid *Lepanthes caritensis* (Tremblay 1997) the stable-stage distribution was skewed towards adults.

Populations that are far from the stable-stage distribution will exhibit short-term behaviour that may differ considerably from the asymptotic behaviour ruled by the dominant eigenvalue. (Horvitz & Schemske 1995, Caswell 2001). The damping ratios were close to or above 1.5 for all species. This means that the dominant eigenvalue was about 50 % larger than the subdominant eigenvalue, thus the convergence to the stable-stage distribution will not be disrupted much by the transient dynamics (Horvitz & Schemske 1995). For *Calathea ovidensis*, damping ratios between 1.38 and 2.29 were found (Horvitz & Schemske 1995). The damping ratio of a population of *Laelia speciosa* was 1.74, correspondingly, it would take ca. 4.1 years for λ_1 to dominate by ten-fold (Hernández-Apolinar), which is in the range found in our study.

In general, mortality decreased with age, which is a common pattern in perennial plants (e. g. Harper 1977, Horvitz & Schemske 1995, Valverde & Silvertown 1998, Godínez-Alvarez & Valiente-Banuet 2004). In the epiphytic orchid *Dimerandra emarginata*, mortality decreased steadily with plant size (Zotz 1998). Seedling mortality in epiphytic bromeliads was high (0.82 – 0.93) in germination experiments conducted by Winkler *et al.* (2005b, see Chapter 3). In *Tillandsia* species monitored using repeated photographs at the same study site, mortality from factors other than branchfall decreased from 33 % for plants <2 cm to ca. 6 % for plants >15 cm (Hietz 1997). Unexpectedly, the clonal species *Jacquiniella teretifolia* showed high adult mortality, which is a consequence of one large branch supporting many reproductive individuals that fell between February 2002 and February 2003).

Average asymptotic population growth rates were below unity in all species, indicating that the populations are declining and, assuming constant environment, will go extinct. However, temporal variability in lambdas of *J. teretifolia* and *L. aromatica* suggests the existence of "good" and "bad" years, in as observed in non-epiphytic plant species (e. g. Aberg 1992, Hoffmann 1999, Pascarella

& Horvitz 1998) and for *Tillandsia brachycaulos* (Mondragón et al. 2004). In this species, population growth rate was correlated with the amount of August precipitation. However, with the current probability of occurrence of "good" years, the population is expected to decline (Mondragón et al. 2004). Detailed climate data for Jalapa up to February 2005 are not available yet. Nevertheless, stochastic population behaviour of the orchids in relation to climate conditions should be considered to make reliable predictions on population development. The only species with an occasional population growth rate equal or greater to one was *J. teretifolia* which occurs frequently in shadow trees of coffee plantations and more open vegetation types and is relatively drought resistant. Hernández-Apolinar (1992) found the population growth rate of *Laelia speciosa* to be above unity on each of three sites, but did not present confidence intervals. Population growth rates for *Lepanthes caritensis* were 0.995 and 0.999 on two trees (Tremblay 1997).

Elasticity analysis

Epiphyte population growth rate was influenced mainly by the survival of pre-reproductive and fertile plants. In *L. aromatica*, the survival of extra-large reproductives is decisive for population maintenance. To illustrate the effect of large elasticity values in a single vital rate, we will give the following example: During the projection interval 2002 – 2003, two reproductive individuals (= 0.6 % of the individuals) fell with their branch leading to a decrease of lambda by 1.8 %. In contrast, when 22 % of the individuals died between 2003 and 2004, lambda decreased only by 4.8 % compared to the previous year because most of the individuals lost were seedlings and infants. Population growth rate of *Laelia speciosa* was influenced most by stasis of juveniles and small reproductives (Hernández-Apolinar 1992) whereas in *Lepanthes caritensis*, the survival of pre-reproductive adults had the greatest influence (Tremblay 1997). In *Tillandsia brachycaulos*, lambda depended mainly on the growth of vegetatively recruited plants, followed by the recruitment of new offshoots (Mondragón et al. 2004).

Fecundity had next to no influence on population growth rate. Thus, herbivory in flowers and fruits of epiphytes, which resulted in the loss of up to 90 % of fruits of *L. aromatica* (Winkler et al. 2005a, Chapter 2) should not have a great impact on population persistence. To obtain a population growth rate greater than one for the transition matrix 2002-2003 (observed lambda: 0.975), fecundity values must be five times the observed values.

Influence of branch characteristics

Small scale population dynamics of epiphytes were influenced by several branch properties. In particular, branch size is related to the probability of an individual to fall with its supporting branch (Hietz 1997). Species with a clumped distribution were more affected by stochastic branchfall, especially if they prefer thinner branches like *Jacquiniella teretifolia*.

Light seems to be the limiting resource for an epiphyte to become fertile and survival probabilities increased with the amount of light (as measured by canopy openness and relative height). This confirms the findings of Winkler *et al.* (2005 b, Chapter 3) who observed that more experimentally exposed bromeliad seedlings survived when canopy openness was high. Nevertheless, this is somewhat surprising, because drought has been reported to be the main cause of death during the early stages of epiphytes (Benzing 1981, Larson 1992).

Bryophytes, lichens and other vascular plants may be competitors (particularly for seedlings), but also provide rooting substrate and maintain a certain humidity level, or – in orchids – provide conditions for the mycorrhizal fungi required for growth and germination. Whether bryophyte and plant cover of branches influenced survival probabilities positively or negatively depended on the season and on the epiphyte species.

Conclusions

The populations of the epiphytes in our study clearly rely on the survival of established plants. Nevertheless, most of the populations studied are declining. In combination with the ever decreasing amount of natural habitats this may lead to the extinction of some of the species. Especially those restricted to closed forests may be seriously threatened. Since nearly all species rely on the survival of large individuals, the frequently occurring extraction of flowering plants for ornamental purposes may be fatal for local populations.

5.5 Acknowledgements

We thank Leticia Cruz Paredes and Angélica Jímenez Aguilar for help in the field. We are grateful to the Instituto de Ecología in Xalapa and to José García Franco for general support. This research was funded by the Austrian Science Fund (FWF grant number P14775).

5.6 Literature

Aberg P. 1992. Size-based demography of the seaweed *Ascophyllum nodosum* in stochastic environments. *Ecology* **73**: 1488-1501.

Agresti A. 2002. *Categorical Data Analysis*. 2nd edition. New York: Wiley-Interscience.

Anonymous. 1999. *S-PLUS 2000. Guide to Statistics, Volume 2*. Seattle: Data Analysis Products Division, MathSoft.

Begon M., Harper J. L. & Townsend C. R. 1990. *Ecology: Individuals, Populations and Communities*. 2nd edition. Cambridge, USA: Blackwell.

Bennett B. C. 1988. *A comparison of life history traits in selected epiphytic and saxicolous species of Tillandsia (Bromeliaceae)*. PhD Thesis. Chapel Hill, NC: University of North Carolina.

Bennett B. C. 1991. Comparative biology of neotropical epiphytic and saxicolous *Tillandsia* spp.: Population structure. *Journal of Tropical Ecology* **7**: 361-371.

Benzing D. H. 1978. Germination and early establishment of *Tillandsia circinnata* Schlecht. (Bromeliaceae) on some of its hosts and other supports in southern Florida. *Selbyana* **5**: 95-106.

Benzing D. H. 1981. The population dynamics of *Tillandsia circinnata* (Bromeliaceae): cypress crown colonies in southern Florida. *Selbyana* **5**: 256-263.

Benzing D. H. 1990. *Vascular Epiphytes. General Biology And Related Biota*. Cambridge: Cambridge University Press.

Benzing D. H. 2000. *Bromeliaceae: Profile of an Adaptive Radiation*. Cambridge: Cambridge University Press.

Buchberger G. 2004. *Dreidimensionale Verteilung von epiphytischen Bromelian und Orchideen in einem humiden Bergwald Mexikos*. Diploma Thesis. Universität für Bodenkultur, Vienna.

Burgman M. A., Ferson S., Akcakaya H. R. 1993. *Risk assessment in conservation biology*. New York: Chapman and Hall.

Calvo R. N. 1993. Evolutionary demography of orchids: Intensity and frequency of pollination and the cost of fruiting. *Ecology* **74**: 1033-1042.

Caswell H. 2001. *Matrix Population Models - Construction, Analysis, and Interpretation*. Second Edition. Sunderland: Sinauer.

De Kroon H., Plaisier A., Van Groenendael J. V. & Caswell H. 1986. Elasticity: the relative contribution of demographic parameters to population growth rate. *Ecology* **67**: 1427-1431. Efron, B. and Tibshirani, R. J. (1993). An Introduction to the Bootstrap. San Francisco: Chapman & Hall.

Godínez-Alvarez H. & Valiente-Banuet A. 2004. Demography of the columnar cactus *Neobuxbaumia macrocephala*: a comparative approach using population projection matrices. *Plant Ecology* **174**: 109-118.

Guárdia R., Raventós J. & Caswell H. 2000. Spatial growth and population dynamics of a perennial tussock grass (*Achnatherum calamagrostis*) in a badland area. *Journal of Ecology* **88**: 950-963.

Harper JL. 1977. *Population biology of plants*. New York: Academic Press.

Hernández-Apolinar M. 1992. *Dinámica poblacional de* Laelia speciosa *(H. B. K.) Schltr. (Orchidaceae)*. Tesis de Licenciatura, Universidad Autónoma de México, Mexico.

Hietz P. & Hietz-Seifert U. 1995. Intra- and interspecific relations within an epiphyte community in a Mexican humid Montane forest. *Selbyana* **16**: 135-140.

Hietz P. 1997. Population dynamics of epiphytes in a Mexican humid montane forest. *Journal of Ecology* **85**: 767-775.

Hietz P., Ausserer J., Schindler G. 2002. Growth, maturation and survival of epiphytic bromeliads in a Mexican humid montane forest. *Journal of Tropical Ecology* **18**: 177-191.

Hoffmann W. A. 1999. Fire and population dynamics of woody plants in a neotropical savanna: matrix model projections. *Ecology* **80**: 1354-1369.

Holdridge L. R. 1967. *Life zone ecology.* San José, Costa Rica: Tropical Science Center.

Horvitz C. C. & Schemske D. W. 1995. Spatiotemporal variation in demographic transitions of a tropical understory herb: projection matrix analysis. *Ecological Monographs* **65**: 155-192.

Larson R. J. 1992. Population dynamics of *Encyclia tampensis* in Florida. - *Selbyana*, **13**: 50-56.

Lefkovitch L. P. 1965. The study of population growth in organisms grouped by stages. *Biometrics* **21**: 1-18.

Menges E. S. 1990. Population viability analysis for an endangered plant. *Conservation Biology* **4**: 52-62.

Mills L. S., Hayes S. G., Baldwin C., Wisdom M. J., Citta J., Mattson D. J. & Murphy K. 1996. Factors leading to different viability predictions for a grizzly bear data set. *Conservation Biology* **10**: 863-873.

Mondragón D., Durán R., Ramírez I. & Valverde T. 2004. Temporal variation in the demography of the clonal epiphyte *Tillandsia brachycaulos* (Bromeliaceae) in the Yucatán Peninsula, Mexico. *Journal of Tropical Ecology* **20**: 189-200.

Pascarella J. B. & Horvitz C. C. 1998. Hurricane disturbance and the population dynamics of a tropical understory shrub: megamatrix elasticity analysis. *Ecology* **79**: 547-563.

Rzedowski J. 1986. *Vegetación de México, 3rd edition.* Mexico: Editorial Limusa.

Silvertown J., Franco M., Pisanty I. & Mendoza A. 1993. Comparative plant demography – relative importance of life-cycle components to the finite rate of increase in woody and herbaceous perennials. *Journal of Ecology* **81**: 465-467.

Tremblay R. L. & Ackerman J. D. 2001. Gene flow and effective population size in *Lepanthes* (Orchidaceae): a case for genetic drift. *Biological Journal of the Linnean Society* **72**: 47-62.

Tremblay R. L. 1997. *Lepanthes caritensis*, an endangered orchid: No sex, no future? *Selbyana* **18**: 160-166.

Valverde T. & Silvertown J. 1997. A metapopulation model for *Primula vulgaris*, a temperate forest understorey herb. *Journal of Ecology* **85**: 193-210.

Valverde T. & Silvertown J. 1998. Variation in the demography of a woodland understorey herb (*Primula vulgaris*) along the forest regeneration cycle: projection matrix analysis. *Journal of Ecology* **86**: 545-562.

Williams-Linera G. 1997. Phenology of deciduous and broadleaved-evergreen tree species in a Mexican tropical lower montane forest. *Global Ecology and Biogeography Letters* **6**: 115-127.

Winkler M, Hülber K, Hietz P. 2005. Effect of canopy position on germination and seedling survival of epiphytic bromeliads in a Mexican humid montane forest. *Annals of Botany* **95**: 1039-1047.

Winkler M. & Hietz P. 2001. Population structure of three epiphytic orchids (*Lycaste aromatica*, *Jacquiniella leucomelana*, and *J. teretifolia*) in a Mexican humid montane forest. *Selbyana* **22**: 27-33.

Winkler M. 2001. Demographie der epiphytischen Orchideen *Lycaste aromatica*, *Jacquiniella leucomelana* und *J. teretifolia*. Diploma thesis. University of Vienna, Vienna.

Winkler M., Hülber K., Mehltreter K., García-Franco J. & Hietz P. 2005a. Herbivory in epiphytic bromeliads, orchids and ferns in a Mexican montane forest. *Journal of Tropical Ecology* **21**: 147-154.

Wisdom M. J., Mills L. S. & Doak D. F. 2000. Life stage simulation analysis: estimating vital-rate effects on population growth for conservation. *Ecology* **81**: 628-641.

Zotz G. 1998. Demography of the epiphytic orchid, *Dimerandra emarginata*. *Journal of Tropical Ecology* **14**: 725-741.

5.7 Appendix

Table 5.6: Transition matrices and demographic properties of the populations. Stages as in Table 5.1, other symbols are q_x = mortality rate, v = stage-specific reproductive value, w = stable-stage distribution, t_d = stage-specific time to death (years).

Feb. 2002-Feb. 2003		s	i	j	p	f or $f1$	$f2$	$f3$	v	w	t_d
Jacquiniella leucomelana											
	s	0.346		0.019	0.007	0.500	0.669	1.318	1.000	0.278	3.59
	j	0.274		0.468	0.014	0.016	0.030	0.000	1.794	0.201	4.61
	p	0.016		0.305	0.489	0.355	0.403	0.368	2.130	0.339	4.15
	$f1$	0.000		0.026	0.113	0.290	0.075	0.000	2.992	0.080	4.50
	$f2$	0.003		0.000	0.064	0.097	0.269	0.015	4.092	0.050	5.20
	$f3$	0.000		0.000	0.035	0.016	0.119	0.522	5.976	0.053	5.45
	q_x	0.361		0.182	0.277	0.210	0.104	0.096			
J. teretifolia											
	s	0.152		0.000	0.000	0.105	0.186	0.551	1.000	0.000	2.95
	j	0.273		0.733	0.011	0.059	0.000	0.000	2.565	0.084	5.50
	p	0.000		0.068	0.846	0.529	0.458	0.176	4.499	0.916	6.89
	$f1$	0.000		0.000	0.000	0.000	0.000	0.000	3.319	0.000	5.26
	$f2$	0.000		0.000	0.000	0.059	0.125	0.000	3.265	0.000	4.93
	$f3$	0.000		0.000	0.000	0.000	0.042	0.412	3.050	0.000	3.76
	q_x	0.576		0.199	0.143	0.353	0.375	0.412			
Lycaste aromatica											
	s	0.346	0.000	0.000	0.000	0.036	0.059	0.102	1.000	0.090	6.41
	i	0.231	0.702	0.033	0.000	0.000	0.000	0.000	2.725	0.078	13.83
	j	0.000	0.137	0.300	0.000	0.029	0.000	0.000	5.020	0.017	21.09
	p	0.000	0.000	0.400	0.486	0.265	0.097	0.154	7.039	0.202	28.39
	$f1$	0.000	0.008	0.067	0.086	0.324	0.000	0.000	7.244	0.029	29.09
	$f2$	0.000	0.000	0.000	0.214	0.324	0.419	0.000	7.564	0.095	29.81
	$f3$	0.000	0.000	0.000	0.157	0.029	0.452	0.821	7.673	0.489	30.01
	q_x	0.423	0.153	0.200	0.057	0.029	0.032	0.026			

Table 5.6 cont.

Feb. 2003-Feb. 2004		s	i	j	p	f(1)	f2	f3	v	w	t_d
Jacquiniella leucomelana											
	s	0.453		0.166	0.009	0.298	0.322	0.612	1.000	0.409	2.18
	j	0.042		0.438	0.077	0.140	0.000	0.000	7.702	0.113	4.16
	p	0.000		0.112	0.243	0.186	0.194	0.105	13.585	0.126	4.83
	f1	0.000		0.071	0.176	0.395	0.139	0.012	18.299	0.122	6.16
	f2	0.000		0.000	0.108	0.163	0.556	0.105	22.096	0.154	6.61
	f3	0.003		0.000	0.104	0.023	0.028	0.547	16.423	0.077	5.02
	q_x	0.503		0.213	0.284	0.070	0.083	0.233			
J. teretifolia											
	s	0.357		0.000	0.000	0.000	4.455	2.227	1.000	0.349	
	j	0.571		0.706	0.029	0.000	0.000	0.000	1.518	0.392	
	p	0.000		0.184	0.699	0.000	0.000	0.000	4.270	0.138	
	f1	0.000		0.007	0.049	0.000	0.000	0.000	0.000	0.008	
	f2	0.000		0.000	0.107	0.000	0.250	0.071	12.769	0.022	
	f3	0.000		0.000	0.078	0.000	0.750	0.929	10.636	0.092	
	q_x	0.071		0.103	0.039	0.000	0.000	0.000			
Lycaste aromatica											
	s	0.136	0.009	0.000	0.000	0.016	0.042	0.102	1.000	0.096	1.65
	i	0.091	0.582	0.000	0.000	0.000	0.000	0.000	8.711	0.025	4.65
	j	0.000	0.109	0.190	0.000	0.000	0.000	0.000	27.620	0.004	8.53
	p	0.000	0.000	0.333	0.603	0.452	0.216	0.041	37.000	0.117	10.79
	f1	0.000	0.000	0.190	0.034	0.194	0.000	0.000	42.531	0.006	12.20
	f2	0.000	0.000	0.000	0.086	0.226	0.270	0.000	48.997	0.017	13.45
	f3	0.000	0.000	0.000	0.121	0.065	0.459	0.898	52.806	0.735	14.14
	q_x	0.773	0.300	0.286	0.155	0.065	0.054	0.061			

Table 5.6 cont.

Feb. 2004 – Feb. 2005		s	i	j	p	f1	f2	f3	v	w	t_d
Jacquiniella leucomelana											
	s	0.257		0.018	0.010	0.269	0.472	0.758	1.000	0.291	3.55
	j	0.268		0.198	0.031	0.014	0.000	0.000	1.516	0.121	4.10
	p	0.112		0.333	0.333	0.347	0.100	0.123	1.997	0.236	4.36
	f1	0.011		0.162	0.167	0.250	0.033	0.014	2.301	0.102	4.52
	f2	0.000		0.009	0.188	0.181	0.267	0.027	2.989	0.105	4.25
	f3	0.000		0.000	0.052	0.000	0.400	0.534	2.952	0.144	3.68
	q_x	0.353		0.279	0.219	0.194	0.183	0.301			
J. teretifolia											
	s	0.559		0.052	0.000	0.223	0.618	3.568	1.000	0.147	7.23
	j	0.322		0.765	0.052	0.000	0.000	0.000	1.018	0.512	6.79
	p	0.000		0.026	0.763	0.667	0.231	0.167	2.776	0.291	8.50
	f1	0.000		0.000	0.041	0.333	0.077	0.000	3.745	0.025	10.00
	f2	0.000		0.000	0.031	0.000	0.538	0.167	4.435	0.026	8.08
	f3	0.000		0.000	0.000	0.000	0.000	0.625	18.220	0.000	10.05
	q_x	0.119		0.157	0.113	0.000	0.154	0.042			
Lycaste aromatica											
	s	0.188	0.000	0.000	0.000	0.020	0.022	0.051	1.000	0.041	23.98
	i	0.563	0.739	0.000	0.000	0.000	0.000	0.000	1.437	0.091	32.81
	j	0.000	0.145	0.250	0.000	0.000	0.000	0.000	2.274	0.018	46.56
	p	0.000	0.014	0.500	0.538	0.111	0.077	0.083	2.902	0.149	58.03
	f1	0.000	0.000	0.100	0.038	0.389	0.000	0.000	2.467	0.012	49.07
	f2	0.000	0.000	0.000	0.250	0.389	0.346	0.000	2.975	0.064	57.94
	f3	0.000	0.000	0.000	0.135	0.000	0.462	0.917	3.659	0.625	70.08
	q_x	0.250	0.101	0.150	0.038	0.111	0.115	0.000			

6 Population dynamics of epiphytic bromeliads: Life strategies and the role of host branches

Manuela Winkler, Karl Hülber and Peter Hietz

published in Basic and Applied Ecology (2007) 8: 183-196.

© 2006 Gesellschaft für Ökologie. Published by Elsevier GmbH (reprinted with permission)

Abstract

Epiphytes need to balance the trade-off between long-term survival and colonizing new branches in their resource-limited and high-risk canopy habitats. We censused populations of five bromeliad species over two years and calculated matrix population models to determine the key factors influencing population growth rates. Additionally, logistic regression models were applied to study how demographic features change with microsite characteristics and disturbance.

The bromeliads studied are slow-growing, with species preferring thinner branches maturing faster. Population growth rates were below unity in all species except for the drought-resistant *Tillandsia juncea*. Population growth rates depended almost exclusively on survival, above all of adult plants. Fecundity had only little impact, even in species of the outer canopy where branchfall-related disturbance and mortality are frequent.

Survival rates and the probability to become reproductive increased with light availability in most species. Microsite characteristics had the greatest impact on seedling survival, although this contributed very little to population growth rates.

We conclude that branchfall-related mortality is a key factor for population persistence of epiphytes dwelling in the outer canopy, and that resource availability constrains the possibility to counteract disturbance with higher fecundity. For species of the outer canopy an increased disturbance by more frequent strong winds could further constrain population growth. Because population persistence strongly relies on the survival of adult plants, harvesting bromeliads for ornamental or other purposes should be restricted to immature individuals. In the future, better models of epiphyte populations in a habitat with highly patchily dynamics should be based on a metapopulation approach.

Key words: Bromeliaceae, *Catopsis*, disturbance, elasticity, epiphytes, matrix population models, Mexico, *Tillandsia*, vital rates

6.1 Introduction

In terrestrial plants three life strategies are commonly distinguished (Grime 1977). Competitors prefer low-stress environments with low disturbance rates and have high growth rates. Ruderals or pioneers have to account for frequent disturbance of their habitat and thus invest overproportionally in reproduction, whereas stresstolerators have to deal with a harsh environment which limits growth rates and leads to late reproduction and slow life cycles. On the one hand, forest canopies are high-risk habitats for epiphytes with high rates of patch turnover (Benzing 1990; Hietz 1997), which should favour investment in high fecundity. On the other hand, reproductive power is reduced in many epiphytes that need to invest in stress tolerance (Benzing 1990). Epiphytes have been shown to be generally slowgrowing and long-lived plants (Benzing, 1981; Hietz *et al.* 2002; Schmidt & Zotz 2002), with the notable exception of twig epiphytes, which are able to complete their life cycles within less than 2 years (Chase 1987; Warford 1992).

Silvertown *et al.* (1993) and Franco & Silvertown (2004) proposed a comparative demography of plants based upon elasticities of vital rates derived from matrix population models. Their demographic triangle composed of survival, growth and fecundity elasticities delimits elasticity space and provides a tool to link plant life history and demography. Matrix population models are also well-established to study aspects of species conservation (Burgman *et al.* 1993; Menges, 2000 and references therein; Wiegand *et al.* 2005; Wisdom *et al.* 2000), spatiotemporal variation and metapopulation biology (Jongejans & De Kroon, 2005; Valverde & Silvertown, 1997, 1998), plant–animal interactions (Horvitz *et al.* 2005; Leimu & Lehtilä 2006), and the influence of environmental stochasticity on populations (Aberg 1992; for an overview see Caswell 2001 and references therein; Hoffmann 1999; Pascarella & Horvitz 1998).

Among all major life forms of tropical plants, the population dynamics of epiphytes are probably the least studied. Apart from more descriptive studies of epiphyte populations (Bennett 1988, 1991; Benzing 1978, 1981; Hietz 1997; Larson 1992; Zotz 1998), several more comprehensive studies have applied matrix models for epiphytes. Calvo (1993) investigated the trade-off between current reproduction and future growth and reproduction in an epiphytic orchid. Effective population sizes of populations of *Lepanthes* orchids were estimated using a matrix-model (Tremblay & Ackerman 2001). Hernández-Apolinar (1992), and Tremblay (1997) used Lefkovitch matrices to describe populations of the orchids *Laelia speciosa* and *Lepanthes caritensis*, respectively. Mondragón *et al.* (2004) used a 3-year demographic data set for the bromeliad *Tillandsia brachycaulos* to project its long-term population dynamics. For the bromeliad *Werauhia sanguinolenta*, the demographic behaviour of populations on three sites with differing moisture input was compared (Zotz 2005), and in a long-term study Zotz et al. (2005) correlated vital rates with the amount of annual precipitation and estimated the impact of epiphyte density and herbivores on population performance. Zotz & Schmidt (2006) analysed possible reasons for the observed population decline of the orchid *Aspasia principissa* in a 7-year study.

The position in the canopy affects epiphytes in many ways. Hietz et al. (2002) showed that the time to reach fertility was positively correlated with the preferred branch diameter of different bromeliad species, and the proportion of reproductive plants was found to depend on branch height and diameter in two orchid species (Winkler & Hietz 2001). Species growing preferentially on thinner branches experience higher mortality because of their insecure support (Hietz 1997). Substrate instability does not only affect epiphyte mortality but can also have a greater impact on population growth rate than drought, because drought kills mainly small plants whereas branchfall is the main cause of mortality in large plants, which have a higher influence on population growth rates (Zotz et al. 2005). However, if death due to branchfall is excluded, *Tillandsia* species experience higher mortality on thick branches, which tend to be in the more humid and shaded zones of the canopy (Hietz 1997). Site-specific mortality was also found in experimentally sown seedlings of epiphytic bromeliads (Winkler et al. 2005a; Zotz & Vollrath 2002).

In this study, we present matrix population models for five abundant species of epiphytic bromeliads. We expected population growth rates of species preferring stable substrate to be near unity and determined by survival. For those on thinner branches we expected population growth to deviate stronger from unity because of the strong stochasticity, and to be determined by fecundity to balance the high death rate caused by branchfall. The main questions were: (1) Which are the key demographic processes determining population growth rates? (2) Is there a relationship between disturbance regime, population growth rates and elasticities? (3) How are epiphyte populations affected by canopy microsites?

6.2 Material and Methods

Study area and species

This study was conducted in the forest reserve 'Santuario Bosque de Niebla' adjacent to the Instituto de Ecología, 2.5 km south of Xalapa, in central Veracruz, Mexico (19°31'N, 96°57'W;

1350m a.s.l.). Average temperature is 19 °C, and annual precipitation is ca. 1500 mm, most of which falls in the rainy season (June–October). The forest is at the transition between premontane and lower montane moist forest according to the Holdridge (1967) life-zone system. In Mexico, it is commonly classified as 'bosque mesófilo de montaña' (mesophilous montane forest) following Rzedowski (1986). Descriptions of the forest structure are given by Williams-Linera (1997) and those of the epiphyte community by Hietz & Hietz-Seifert (1995).

We censused populations of the epiphytic bromeliads *Catopsis sessiliflora* (Ruiz and Pav.) Mez, *Tillandsia deppeana* Steud., *T. juncea* (Ruiz and Pav.) Poir., *T. multicaulis* Steud., and *T. punctulata* Schltdl. and Cham. All *Tillandsia* juveniles and adults of *T. juncea* exhibit the atmospheric habit meaning that they possess narrow leaves that do not hold water, but display confluent layers of absorbing trichomes. *C. sessiliflora* of all sizes and intermediate to adult *T. multicaulis* and *T. deppeana* have broad thin leaves that form tanks, and *T. punctulata* is a tank-atmospheric intermediate species. *T. deppeana* is largely monocarpic, the other species are polycarpic. Except for the CAM-plant *T. juncea*, all species exhibit C3-photosynthesis.

Population census

In August 2001, 180 branch sections with bromeliads growing on them were selected and marked. Branch sections were 10–250 cm long and distributed throughout the crowns of nine trees with a maximum distance between two trees of ca. 250 m. All individuals of the study species on these branch sections were tagged, the number of vegetative and reproductive ramets was recorded and the longest leaf was measured to the nearest 0.5 cm.

Branch height relative to total tree height and distance to the trunk relative to the crown radius, inclination, circumference, and canopy openness (using a convex spherical densiometer, Ben Meadows, Atlanta, GA), were measured and the proportion of branch surface covered by bryophytes and total epiphyte cover (*i.e.*, vascular plants, bryophytes and lichens) were estimated for each branch section. Canopy openness, relative height and relative distance were used as surrogates of light availability. In the rainy season, the latter two may be more representative of light because when all trees are fully in leaf, differences in canopy openness are not easily detected with a densiometer. Brypophyte cover is a rough measure of humidity, and branch diameter of substrate roughness as large branches have a thicker and more corrugated bark. Survival, vegetative growth and reproduction of the tagged individuals were monitored and newly emerging seedlings were recorded. Populations were censused in February and August 2002, in February 2003 and parts of the populations again in February 2004.

Matrix population models

To characterize population dynamics we used Lefkovitch (1965) stage-classified transition matrices. A combination of size (length of longest leaf) and reproductive criteria was used to distinguish stage classes (as in Menges 1990; Horvitz & Schemske 1995; Table 6.1). The upper size threshold for seedlings was set at ≤1.0 cm which corresponds to the 95th percentile size of 1-year-old seedlings from germination experiments (Winkler *et al.* 2005a and unpubl. data). Non-reproductive adults were defined as non-flowering plants at least the size of the 5th percentile of the size distribution of reproductive plants. The limit between infants and juveniles is the mid-interval between the upper size threshold of seedlings and the lower size threshold of non-reproductive adults.

Table 6.1: Definition of stage classes according to the size of the longest leaf (cm) and reproductive status.

Species	Seedlings s	Infants i	Juveniles j	Non-reproductive adults	Reproductive (flowering) adults
Catopsis sessiliflora	≤1.0	1.0 - 7.0	7.0 - 13.0	>13.0	any size
Tillandsia deppeana	≤1.0	1.0 - 18.0	18.0 - 34.5	>34.5	"
T. juncea	≤1.0	1.0 - 12.5	12.5 - 24.0	>24.0	"
T. mulitcaulis	≤1.0	1.0 - 11.5	11.5 - 21.5	>21.5	"
T. punctulata	≤1.0	1.0 - 13.5	13.5 -25.5	>25.5	"

Fecundity was calculated as the number of newly emerged seedlings at the end of a projection interval (1 year) divided by the number of reproductive individuals at the beginning of the interval. Although seedlings, broad-leaved infants and juveniles of *Tillandsia* are difficult to distinguish, ca. 2/3 of the seedlings could be assigned to a species because they mass-germinated close to the mother plant. Transition probabilities were first calculated separately for identified seedlings, as well as for unidentified seedlings, broad-leaved infants, and juveniles. For the matrix models, the transition probabilities of seedlings of a given species were then calculated by including all identified plants plus a fraction of the unidentified plants equal to the species' estimated contribution to the *Tillandsia* seed rain. Seed rain was calculated as the number of reproductive individuals x the average number of seeds per individual. Reproductive individuals in the forest were mapped by Buchberger (2004), and seed numbers were counted from 31 to 34 capsules per species (unpubl. data). The transition probabilities of broad-leaved infants and juveniles were calculated similarly, assigning probabilities of unidentified plants according to the proportion of *T. deppeana* and *T. multicaulis* recorded on photographed branches (Hietz 1997).

A 5x5 transition matrix was built for each of the two projection intervals studied (February 2002–2003, February 2003–2004). These transition matrices were summed elementwise to create a pooled matrix for each species (compare Horvitz & Schemske 1995). The population growth rate λ (the dominant eigenvalue of the transition matrix), the stage-specific reproductive value v (the left eigenvector) and the stable-stage distribution w (the right eigenvector, both scaled so that the vector's sum is 1) were calculated using the eigen.Matrix command from the S-Plus Matrix library (S-Plus 2000 MathSoft Inc. 1988–1999). Empirical confidence intervals for λ were obtained using non-parametric bootstrapping (*bootstrap* command in S-Plus 2000) with 1000 resamples drawn (Caswell 2001; Efron & Tibshirani 1993). The stable-stage distribution gives the proportion of individuals in each stage to which the relative frequencies of stages will converge regardless of initial frequencies (Caswell 2001; Horvitz & Schemske 1995). The damping ratio ($p = \lambda_1 / |\lambda_2|$, where λ_2 is the second largest eigenvalue) was used as a measure of the time to reach the stable-stage distribution. A G-test was performed to test for differences between observed and stable-stage distribution using Pop Tools (Hood 2004).

Subtracting the transition matrix from the inverse of its identity matrix gives a matrix representing the mean time spent in each stage (Caswell 2001). Adding the entries on its main diagonal except that for the first reproductive stage gives the time to first reproduction (t_{rep}), and its column sums give the mean time to death (*i.e.*, the remaining life time) for each stage (t_d).

For each transition matrix, an elasticity matrix was constructed with the elasticity e_{ij} of λ defined as

$$e_{ij} = \left(\frac{a_{ij}}{\lambda_1} \frac{\delta \lambda}{\delta a_{ij}} \right) = \frac{1}{\lambda_1} s_{ij} a_{ij}$$

where a_{ij} is an entry in the transition matrix and s_{ij} is the absolute sensitivity of λ to changes in a_{ij} (Caswell 2001; De Kroon *et al.* 1986). The elasticity value for a particular matrix element (e_{ij}) measures the effect of changes in a_{ij} on λ (Caswell 2001). Elasticity values can be summed up across vital rates (survival, growth, fecundity), but since every e_{ij} is a function of more than one of these vital rates (e.g., positive growth implies survival), elasticities for survival, positive growth, negative growth, and fecundity were calculated using the formulae summarised in Franco and Silvertown (2004, Eqs. (1)–(4) and (7)–(10)) to estimate the contributions of each vital rate to λ. The obtained elasticity values were standardized so that their sum equals one (Franco & Silvertown 2004).

Logistic regression models

To evaluate the influence of branch properties on the probability of growth from the non-reproductive adult to the reproductive stage we used simple logistic regression models. Regression coefficients were standardized by dividing the regression coefficients by their standard error to achieve comparability among variables with different units of measurement (Tabachnick & Fidell 1996). Proportional odds models were used to test the influence of single branch properties on stage-specific mortality. These are cumulative logit models simultaneously using all response categories in a single model. Each cumulative logit has its own intercept (α_j) increasing with j, but coefficients are the same (β):

logit $[P(y \leq j \mid x)] = \alpha_j + \beta' x$

where $j = 1, \ldots, J\text{-}1$ is the respective stage (Table 6.1), and x is the branch property (Agresti 2002). Somer's D_{xy} (ranging from 0 to 1) measures the predictive value of a model. Logistic regression models were analysed using the lrm command of the S-Plus Design-library (Anonymous 1999).

6.3 Results

Demographic properties

Apart from stasis and progressive development, the life cycles of the study species also show retrogression by one or two stages (*i.e.*, individuals can decrease in size, for example, when the leading shoot dies; Table 6.2). Reproductive values increased markedly with plant size in all species, *i.e.*, seedlings contribute only little to reproduction. Time to reproduction obtained from the pooled transition matrices ranged from ca. 9 years in *C. sessiliflora* and *T. deppeana* to 15–16 years in *T. juncea* and *T. multicaulis* (Table 6.3).

Table 6.2: Pooled transition matrices and demographic properties of epiphyte populations.

		s	i	j	a	f	$v[\%]$	$w[\%]$	$osd[\%]$	n	t_d
Catopsis sessiliflora											
	s	0.147	0.014	0.000	0.000	4.429	0.6	36.0	23.2	29	1.90
	i	0.235	0.571	0.040	0.000	0.000	1.6	35.1	39.2	49	2.64
	j	0.000	0.029	0.520	0.120	0.000	13.7	9.1	16.8	21	3.65
	a	0.000	0.000	0.120	0.480	0.714	34.1	14.4	16.0	20	5.37
	f	0.000	0.000	0.000	0.200	0.286	50.0	5.4	4.8	6	6.77
	q_x	0.618	0.386	0.320	0.200	0.000					
Tillandsia deppeana											
	s	0.179	0.010	0.000	0.000	49.444	0.6	32.1	21.6	72	2.39
	i	0.251	0.715	0.049	0.000	0.000	1.5	62.2	68.2	227	3.80
	j	0.000	0.018	0.538	0.042	0.222	10.6	4.1	2.4	8	3.21
	a	0.002	0.000	0.105	0.292	0.444	30.6	1.2	5.7	19	2.83
	f	0.000	0.000	0.000	0.292	0.000	56.7	0.4	2.1	7	2.97
	q_x	0.568	0.257	0.308	0.375	0.333					
T. juncea											
	s	0.093	0.000	0.000	0.000	6.714	0.9	49.4	17.4	25	2.35
	i	0.244	0.618	0.033	0.000	0.000	3.4	30.9	18.1	26	4.65
	j	0.000	0.029	0.567	0.077	0.000	14.3	3.6	20.1	29	10.30
	a	0.000	0.029	0.200	0.631	0.286	31.3	9.3	42.4	61	16.52
	f	0.000	0.000	0.000	0.215	0.714	50.2	6.8	2.1	3	20.02
	q_x	0.664	0.324	0.200	0.077	0.000					
T. multicaulis											
	s	0.123	0.009	0.000	0.007	16.041	0.7	44.6	13.0	58	2.23
	i	0.221	0.704	0.039	0.007	0.000	2.2	39.3	43.3	193	4.21
	j	0.000	0.031	0.390	0.029	0.082	18.0	3.0	7.8	35	7.31
	a	0.002	0.000	0.284	0.730	0.673	36.4	10.8	25.8	115	11.84
	f	0.000	0.000	0.000	0.182	0.102	42.8	2.3	10.1	45	10.65
	q_x	0.655	0.256	0.288	0.044	0.143					
T. punctulata											
	s	0.125	0.000	0.000	0.000	0.519	0.0	6.6	4.9	7	1.46
	i	0.067	0.692	0.000	0.014	0.000	0.0	19.6	22.5	32	4.16
	j	0.000	0.103	0.633	0.099	0.000	0.0	43.6	16.2	23	2.72
	a	0.000	0.000	0.000	0.592	0.407	38.0	22.6	40.8	58	4.37
	f	0.000	0.000	0.000	0.085	0.481	62.0	7.7	15.5	22	5.35
	q_x	0.808	0.205	0.367	0.211	0.111					

Stages as in Table 6.1; other symbols are q_x = mortality rate, v = stage-specific reproductive value, w = stable-stage distribution, n = number of individuals and osd = observed stage distribution February 2002, t_d = stage-specific time to death (years).

The small stages dominated the stable-stage distributions (Table 6.2). The observed stage distribution was dominated by the infant stage in *C. sessiliflora*, *T. deppeana* and *T. multicaulis*, whereas in *T. juncea* and *T. punctulata* nonreproductive adults prevailed. Observed and stable-stage distribution differed significantly in all species except *C. sessiliflora* (G-test: $G = 6.7$, 23.5, 81.6, 143.5 and 30.0 in *C. sessiliflora*, *T. deppeana*, *T. juncea*, *T. multicaulis* and *T. punctulata*, respectively; $df = 4$). The damping ratios were close to or above 1.5 for all species except *Tillandsia punctulata* (Table 6.3), where λ was only a little larger than λ_2. Mortality rates in the seedling stage exceeded 0.5 in all species (Table 6.2). Mortality generally decreased markedly with age, except for the monocarpic *T. deppeana*. Mean time to death (t_d) correspondingly increased with age in most species (Table 6.2). Population growth rates of the pooled transition matrices were <1 in all species except *T. juncea*, and *T. punctulata* when considering the confidence intervals (Table 6.3). Considering single-year transition matrices, only *T. multicaulis* reached a λ of 1.013 in the first year (data not shown).

Table 6.3: Population growth rate (λ, with 5% and 95%-confidence limits derived from bootstrapping) with and without individuals fallen with their supporting branches, percentage of transition events that were deaths attributable to branchfall ($d_{branchfall}$), damping ratio (ρ), and time to reproduction (in years, t_{rep}) obtained from the pooled transition matrices.

	with branchfall		without branchfall				
	λ	(5 – 95 %CL)	λ	(5 – 95 %CL)	$d_{branchfall}$	ρ	t_{rep}
Catopsis sessiliflora	0.82	(0.71-0.92)	0.88	(0.78-0.96)	6.2	1.50	9.31
Tillandsia deppeana	0.85	(0.71-0.98)	0.89	(0.74-1.02)	9.6	1.45	8.88
T. juncea	1.01	(0.93-2.31)	1.03	(0.95-2.31)	2.1	1.60	15.28
T. multicaulis	0.96	(0.92-0.99)	0.98	(0.94-1.01)	2.2	1.67	15.47
T. punctulata	0.73	(0.69-1.12)	0.75	(0.71-1.20)	1.8	1.06	10.04

Elasticity analysis

Survival was the most important demographic process in all species (Figure 6.2A). Stasis of nonreproductive and reproductive adult stages was the vital rate with the highest elasticity values (Figure 6.1). Only for *T. deppeana*, the contribution of the infant stage predominated (Figure 6.2B). The relative importance of fecundity was negligible in all species (Figure 6.2A).

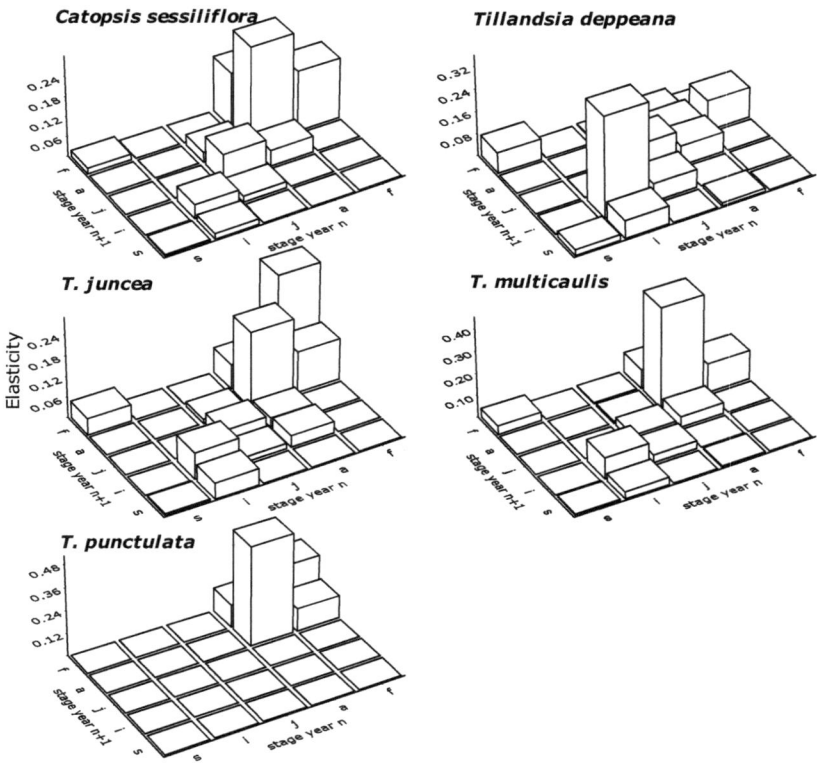

Figure 6.1: Elasticities (e_{ij}) of λ to changes in the entries a_{ij} of the pooled transition matrices. Note the different scales on the Z-axis. s = seedlings, i = infants, j = juveniles, a = non-reproductive adults, f = reproductive adults.

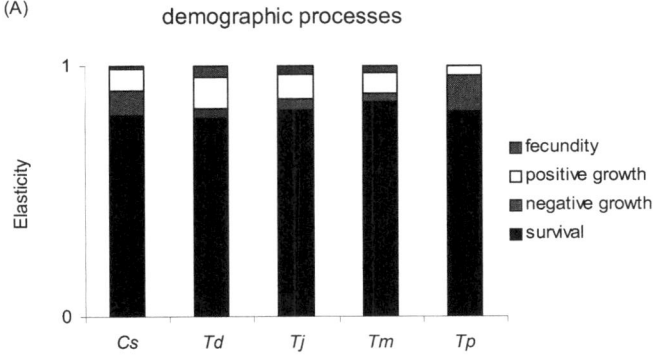

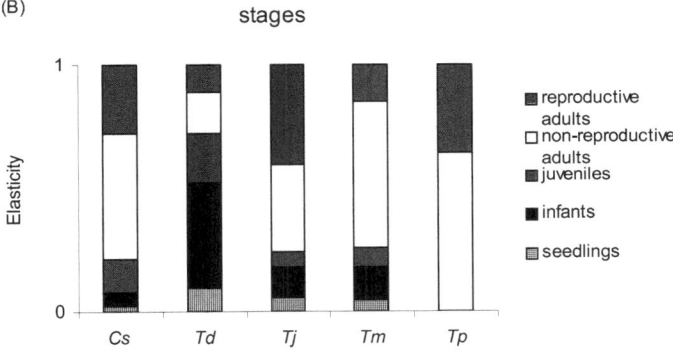

Figure 6.2: Elasticity values of pooled transition matrices for (A) demographic processes (calculated following Franco & Silvertown 2004) and (B) stages. Species: *Cs* = *Catopsis sessiliflora*, *Td* = *Tillandsia deppeana*, *Tj* = *T. juncea*. *Tm* = *T. multicaulis*, *Tp* = *T. punctulata*.

Effects of disturbance and microsite characteristics

Between 1.8% (*T. punctulata*) and 9.6% (*T. deppeana*) of the transition events in the pooled matrices were deaths due to branchfall (Table 6.3). The influence of higher mortality because of branchfall on λ was low and differences were within the range of the confidence intervals. Nevertheless, preferred branch diameter (Buchberger 2004) and population growth rates showed a positive relationship (Figure 6.3). There was also a weak positive relationship between preferred branch diameter and survival elasticities, whereas growth and fecundity elasticities did not vary with branch diameter.

The probability that an individual of a given stage survived an interval between two censuses was influenced by properties of the supporting branch (Figure 6.4). Increased canopy openness or relative height in the tree resulted in high survival probabilities in *Tillandsia* juveniles. For *C. sessiliflora*, higher canopy openness was detrimental in the dry season, but being located higher in the tree was beneficial in the rainy season. A higher plant cover increased survival probabilities of *C. sessilflora* in the dry season, and of *Tillandsia* spp. in the rainy season. There was also a significant positive relationship between survival probability of *Tillandsia* spp. in the rainy season and branch inclination, circumference and bryophyte cover (data not shown). In general, seedlings responded more strongly to environmental conditions than later stages. Similar to survival, the probability that a non-reproductive adult plant becomes reproductive within 1 year increased with canopy openness, relative height and relative distance (Table 6.4). Other branch properties had no effect.

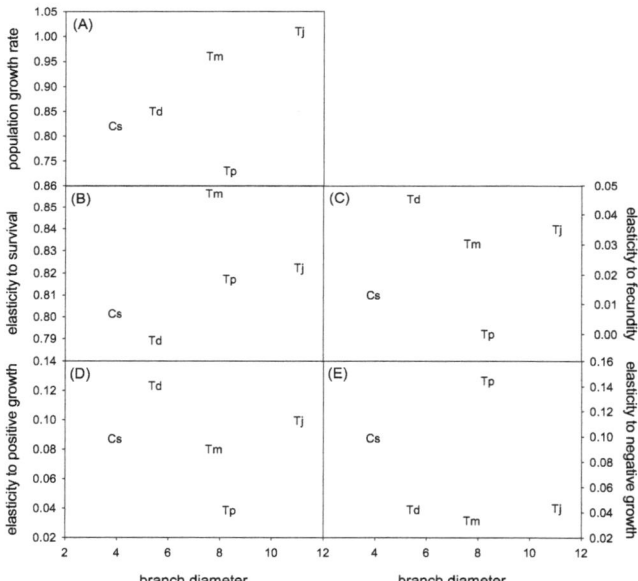

Figure 6.3: Relationship between preferred branch diameter of a species and (A) population growth rates, (B) elasticity of population growth rates to survival, (C) to fecundity, (D) to negative growth and (E) to positive growth. *Cs* = *Catopsis sessiliflora*, *Td* = *Tillandsia deppeana*, *Tj* = *T. juncea*. *Tm* = *T. multicaulis*, *Tp* = *T. punctulata*.

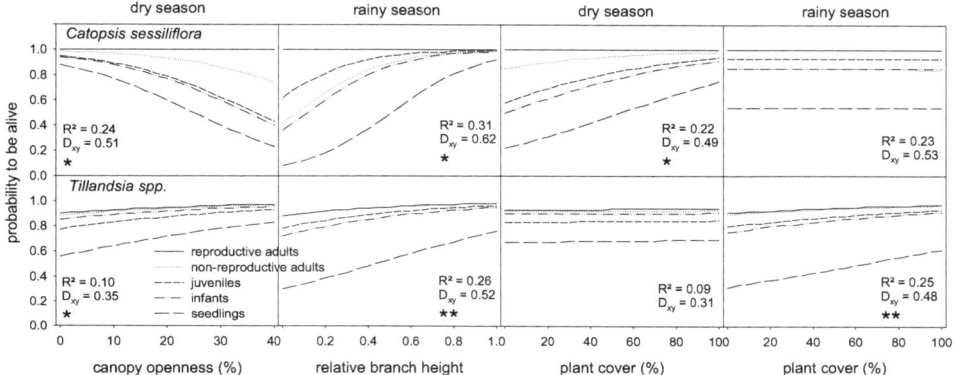

Figure 6.4: Stage-specific survival probability in the intervals February 2002 - August 2002 (dry season) and August 2002 - February 2003 (rainy season), respectively, depending on canopy openness or relative branch height as a measure of light availability, and plant cover. R^2 and Somer's D_{xy} from proportional odds models are given. Individuals fallen with their supporting branch were excluded from the analyses. *Tillandsia* spp. were pooled because seedlings and broad-leaved infants can not be identified to species. The stage classification followed the criteria for *T. multicaulis*, the smallest of the four *Tillandsia* species. Significance levels: * $p<0.05$, ** $p<0.001$.

6.4 Discussion

Demographic patterns

The stable-stage distribution of the populations studied was skewed towards the seedling, infant or juvenile stages. The same is true for the epiphytes *L. speciosa* (Hernández-Apolinar 1992), *T. brachycaulos* (Mondragón et al. 2004) where infant offshoots predominated, and *W. sanguinolenta* (Zotz et al. 2005). This is also typical for many terrestrial herbs in various ecosystems (e.g., Godínez-Alvarez & Valiente-Banuet 2004; Horvitz & Schemske 1995; Valverde & Silvertown 1998), although in the epiphytic orchid *L. caritensis* (Tremblay 1997) the stable-stage distribution is skewed towards adults. Populations that are far from the stable-stage distribution will exhibit a short-term behaviour that may differ considerably from the asymptotic behaviour determined by the population growth rate (Caswell 2001; Horvitz & Schemske 1995). Though most of the species' actual distributions differed significantly from their stable-stage distribution, the observed damping ratios suggest that the populations are close to a stable-stage distribution. However, this was not the case for *T. punctulata* ($\lambda = 1.06$), so conclusions drawn from its transition matrix should be treated with caution, considering that not all *Tillandsia* seedlings could be determined to species level.

The epiphytes of our study are slow-growing and reach fertility only after 9–16 years (Table 6.3). This is close to estimates derived from size measurements on repeated photographs taken between 1992 and 1998, which predicted *C. sessiliflora* and *T. deppeana* to reach fertility after 9 and 11 years, respectively, *T. multicaulis* and *T. punctulata* after 13 years and *T. juncea* after 18 years (Hietz *et al.* 2002). Reproductive values increased with size, with adults accounting for more than 80%. Similarly, atmospheric juveniles of *W. sanguinolenta* have a very low reproductive value compared to the largest tank size class which contribute 440% (Zotz 2005). For the epiphytic bromeliad *T. brachycaulos*, reproductive values of the largest size class account for only 17–24.9%, because of higher juvenile survival (Mondragón *et al.* 2004). Reproduction and survival influence the reproductive value of a stage. The low values of seedlings reflect the high probability that a seedling will die before reaching reproductive size and the long time it takes to grow to this size (Caswell 2001).

Mortality decreased with age in the polycarpic species studied, which is a common pattern in perennial plants (e.g., Godínez-Alvarez & Valiente-Banuet 2004; Harper 1977; Horvitz & Schemske 1995; Valverde & Silvertown 1998). In *Tillandsia* species monitored using repeated photographs at the same study site, mortality from factors other than branchfall decreases from 33% for plants <2 cm to ca. 6% for plants >15 cm (Hietz 1997), and in the epiphytic orchid *Dimerandra emarginata*, mortality decreases steadily with plant size (Zotz 1998). In contrast, Mondragón *et al.* (2004) found mortality rates to be more or less evenly distributed among size classes. However, Mondragón *et al.* defined ramets instead of genets as individuals. This may be justified where vegetative reproduction is common, but in our plants we never observed a separation of ramets into independent and unconnected plants. We observed a similar relatively uniform mortality in monocarpic *T. deppeana*, which produces (mostly) only a single rosette. Seedling mortality in the same bromeliad species we studied was also high (82–93%, compared to 57–81% in this study) in germination experiments conducted by Winkler *et al.* (2005a). In contrast, mortality in the smallest size class of *W. sanguinolenta* is considerably lower than in the seedlings in our study (<35%, Zotz 2005; Zotz *et al.* 2005).

Population growth rates were below unity in most species, suggesting that the populations are declining and, assuming a constant environment, will go extinct in the long run. The only species in our study with a population growth rate ≥1 were *T. juncea* and *T. multicaulis*. The first species occurs frequently on shade trees of coffee plantations and more open vegetation types and is a droughtresistant CAM plant. However, when calculating matrix models for the projection interval August 2001 to August 2002, population growth rates were not significantly different from one in any of the species studied (data not shown). Populations of other epiphytes were found to be growing (Hernández-Apolinar 1992; Zotz 2005; Zotz *et al.* 2005), or shrinking (Mondragón *et al.* 2004; Tremblay 1997; Zotz 2005). There is strong evidence for the existence of "good" and "bad" years for epiphytic (Mondragón *et al.* 2004) as well as non-epiphytic plant species (e.g., Aberg 1992; Hoffmann 1999; Pascarella & Horvitz 1998), but at least in some epiphyte species site variability seems to be more important than temporal variability (e.g., Zotz 2005).

Vital rates determining population dynamics

Growth rates of the populations studied depended almost exclusively on survival, above all of adult plants, followed by growth and, with virtually no effect, fecundity. Thus, in elasticity space these epiphytes behave like their host trees or many iteroparous forest herbs (Franco & Silvertown 2004). The same is true for *W. sanguinolenta* (Zotz 2005; Zotz *et al.* 2005), *A. principissa* (Zotz & Schmidt 2006), *L. caritensis* (Tremblay 1997) and *L. speciosa* (Hernández-Apolinar 1992). Except for *L. speciosa*, the stasis of large plants is the most important transition. In contrast, in *T. brachycaulos*, the population growth rate depends mainly on the growth and recruitment of offshoots (Mondragón *et al.* 2004).

Fecundity had a negligible influence on population growth rates of the bromeliads studied. Thus, herbivory damage of flowers and fruits, which resulted in the loss of ca. 15% in *T. deppeana*, *T. juncea* and *T. punctulata* (Winkler *et al.* 2005b) should not have a great impact on population persistence. To achieve unity, fecundity values would have to be three times the observed values in *T. multicaulis*, six times in *T. deppeana*, and 50 times in *C. sessiliflora*, if the other parameters are kept constant. Calculated fecundity does not contribute at all to the population growth rate in *T. punctulata*. Likewise, the contribution of the seedling stage to population growth rates is small, and a fairly unrealistic seedling survival of next to 100% would be necessary to raise population growth rates to unity. We are therefore confident that the potential error caused by unidentified *Tillandsia* seedlings is negligible. Winkler *et al.* (2005a) suggested that high mortality in bromeliad seedlings might represent a bottleneck for the populations of the study species, but this is not supported by our results. However, it should be taken into account that seedling recruitment is a highly variable process and seedlings of *T. deppeana* and *T. multicaulis* were observed to mass-recruit in favourable years. This may have no effect on transition probabilities and consequently on elasticity values, but it does have an effect on the number of individuals in the population.

Effects of disturbances and microsite characteristics

Small-scale population dynamics of epiphytes are influenced by several branch properties. In particular, branch size is related to the probability of an individual to fall with its supporting branch (Hietz 1997). Species preferring thinner branches reached maturity earlier (compare also Hietz *et al.* 2002) and had lower population growth rates. Nevertheless, populations of epiphytes specialised on thinner branches also depended mainly on survival, but to a somewhat lesser extent than species with a preference for stable branches. However, contrary to our expectations the importance of fecundity was invariably negligible and not related to branch diameter. We therefore conclude that branchfall related mortality is a key factor for population persistence of epiphytes dwelling in the outer canopy and that resource availability constrains the possibility to counteract disturbance with higher fecundity. Although we could not account for important factors such as previous resource allocation to reproduction and biomass, our models yielded surprisingly high predictive values. Light seems to be the limiting resource for many epiphytes to become fertile (Table 6.4).

Survival probabilities increased with the amount of light (Figure 6.4). This confirms earlier observations suggesting that the survival of experimentally exposed bromeliad seedlings increased with canopy openness (Winkler *et al.* 2005a), but is nevertheless somewhat surprising, because drought has been reported to be the main cause of death for the early stages of epiphytes (Benzing 1981; Larson 1992; Zotz & Hietz 2001). Only in *C.sessiliflora* did mortality increase with light availability in the dry season. The same pattern was found in germination experiments for the seedlings of this species, which Winkler *et al.* (2005a) explained with the mesomorphic habit of *C. sessiliflora* seedlings versus the drought-resistant atmospheric ones of *Tillandsia*. Bryophytes, lichens and other vascular plants may be competitors (particularly for seedlings), but also provide rooting substrate and higher substrate humidity. Whether plant cover influenced survival probabilities positively or negatively depended on the season and on the epiphyte species. However, since these microsite conditions affected mostly seedlings and infants, which contribute very little to population growth rates, population persistence does not primarily depend on these conditions.

Table 6.4: Probability that a non-reproductive adult will become reproductive within the projection interval Feb 2002 – Feb 2003 depending on branch parameters.

	Branch parameter						
	canopy openness	relative height	relative distance	circum- ference	incli- nation	bryophyte cover	plant cover
Catopsis sessiliflora							
Stand. Coef.	-0.77	-0.95	**1.96**	-1.20	-1.60	-0.48	0.45
R^2	0.06	0.09	0.64	0.10	0.25	0.03	0.01
D_{xy}	0.06	0.22	0.86	0.24	0.53	0.22	0.07
Tillandsia deppeana							
Stand. Coef.	0.20	0.95	1.77	-0.79	-0.97	-0.70	0.00
R^2	0.19	0.11	0.40	0.09	0.10	0.06	0.00
D_{xy}	0.29	0.06	0.71	0.14	0.34	0.14	0.14
T. juncea							
Stand. Coef.	**2.35**	**2.79**	1.66	**-2.67**	**-2.50**	0.67	1.76
R^2	0.17	0.24	0.07	0.21	0.13	0.02	0.10
D_{xy}	0.39	0.52	0.37	0.49	0.37	0.12	0.35
T. multicaulis							
Stand. Coef.	0.57	0.73	0.27	-2.22	**-2.73**	0.83	0.00
R^2	0.01	0.01	0.00	0.04	0.15	0.01	0.00
D_{xy}	0.54	0.10	0.08	0.18	0.40	0.13	0.05
T. punctulata							
Stand. Coef.	0.27	**1.88**	1.10	**-1.93**	-0.91	-1.21	-0.59
R^2	0.00	0.22	0.07	0.27	0.06	0.09	0.03
D_{xy}	0.13	0.57	0.38	0.73	0.33	0.40	0.22

Standardized coefficients, R^2 and Somer's D_{xy} (a measure of the predictive value) from logistic regression models are given. Coefficients significant at the 0.05 level are in bold. Minus signs indicate negative correlations between survival probability and a branch parameter. Adults that died in the projection interval were excluded from the analyses.

Conclusions

Since nearly all species rely on the survival of adults, the common collection of flowering plants for ornamental purposes may be critical for epiphyte populations. We strongly suggest to collect juvenile or sub-mature individuals only, and harvesting should be accompanied by monitoring the respective populations. Although not affected by extraction four of the five populations we studied appeared to be declining. This may indicate a (climate-change related?) long-term trend. Especially for species of the outer canopy an increase in the disturbance rate brought about by more frequent and stronger tropical storms (Webster *et al.* 2005) could affect population growth. However, the observation period was relatively short and might not be representative. Furthermore, to account for the dynamic character of the epiphytic habitat, observing the same set of branches selected initially may not be an ideal approach. As the substrate and microclimate a branch offers for epiphytes change, the branch will at some point become less suitable for the species that colonized it.

Meanwhile, in other parts of the canopy new branches are produced and grow, whose habitat quality also changes making them suitable for epiphyte colonisation. A metapopulation approach can account for the extinction in and recolonisation of patches and the creation of new ones (Hanski & Gilpin, 1997) and may produce more accurate models of epiphyte population dynamics, especially of those with a preference for thin branches.

6.5 Acknowledgements

We thank Leticia Cruz Paredes and Angélica Jímenez Aguilar for help in the field. Gerald Buchberger kindly provided data on the density of epiphytes. We are indebted to the Instituto de Ecología in Xalapa and to José García Franco for general support. The comments of two anonymous reviewers substantially improved the manuscript. This research was funded by the Austrian Science Fund (FWF Grant number P14775 and P17875).

6.6 Literature

Aberg P. 1992. Size-based demography of the seaweed *Ascophyllum nodosum* in stochastic environments. *Ecology,* **73**: 1488-1501.

Agresti A. 2002. *Categorical Data Analysis.* (2nd ed.). New York: Wiley-Interscience.

Anonymous. 1999. *S-PLUS 2000. Guide to Statistics, Volume 2.* Seattle: Data Analysis Products Division, MathSoft.

Bennett B. C. 1988. *A comparison of life history traits in selected epiphytic and saxicolous species of Tillandsia (Bromeliaceae).* PhD Thesis. Chapel Hill, NC: University of North Carolina.

Bennett B .C. 1991. Comparative biology of neotropical epiphytic and saxicolous *Tillandsia* spp.: Population structure. *Journal of Tropical Ecology* **7**: 361-371.

Benzing D. H. 1978. Germination and early establishment of *Tillandsia circinnata* Schlecht. (Bromeliaceae) on some of its hosts and other supports in southern Florida. *Selbyana* **5**: 95-106.

Benzing D. H. 1981. The population dynamics of *Tillandsia circinnata* (Bromeliaceae): cypress crown colonies in southern Florida. *Selbyana* **5**: 256-263.

Benzing D. H. 1990. *Vascular Epiphytes. General Biology And Related Biota.* Cambridge: Cambridge University Press.

Buchberger G. 2004. *Dreidimensionale Verteilung von epiphytischen Bromelian und Orchideen in einem humiden Bergwald Mexikos.* Diploma Thesis. Vienna: Universität für Bodenkultur.

Burgman M. A., Ferson S. & Akcakaya H. R. 1993. *Risk assessment in conservation biology.* New York: Chapman and Hall.

Calvo R. N. 1993. Evolutionary demography of orchids: Intensity and frequency of pollination and the cost of fruiting. *Ecology* **74**: 1033-1042.

Caswell H. 2001. *Matrix Population Models - Construction, Analysis, and Interpretation.* (2nd ed.). Sunderland: Sinauer.

Chase M. W. 1987. Obligate twig epiphytism in the Oncidiinae and other neotropical orchids. *Selbyana* **10**: 24-30.

De Kroon H., Plaisier A., Van Groenendael J. V. & Caswell H. 1986. Elasticity: the relative contribution of demographic parameters to population growth rate. *Ecology* **67**: 1427-1431.

Efron B. & Tibshirani R. J. 1993. *An Introduction to the Bootstrap.* San Francisco: Chapman & Hall.

Franco M. & Silvertown J. 2004. A comparative demography of plants based upon elasticities of vital rates. *Ecology* **85**: 531-538.

Godínez-Alvarez, H. & Valiente-Banuet, A. 2004. Demography of the columnar cactus *Neobuxbaumia macrocephala*: a comparative approach using population projection matrices. *Plant Ecology* **174**: 109-118.

Grime J. P. 1977. Evidence for the existence of three primary strategies in plants and its relevance to ecological and evolutionary theory. *American Naturalist* **111**: 1169-1194.

Hanski I. A. & Gilpin M. E. (Eds.). 1997. *Metapopulation biology: ecology, genetics and evolution.* London: Academic Press.

Harper J. L. 1977. *Population biology of plants.* New York: Academic Press.

Hernández-Apolinar M. 1992. *Dinámica poblacional de* Laelia speciosa *(H. B. K.) Schltr. (Orchidaceae)*. Tesis de Licenciatura. Mexico : Universidad Autónoma de México.

Hietz P. 1997. Population dynamics of epiphytes in a Mexican humid montane forest. *Journal of Ecology* **85**: 767-775.

Hietz P., Ausserer J. & Schindler G. 2002. Growth, maturation and survival of epiphytic bromeliads in a Mexican humid montane forest. *Journal of Tropical Ecology* **18**: 177-191.

Hietz P. Hietz-Seifert U. 1995. Intra- and interspecific relations within an epiphyte community in a Mexican humid Montane forest. *Selbyana* **16**: 135-140.

Hoffmann W. A. 1999. Fire and population dynamics of woody plants in a neotropical savanna: matrix model projections. *Ecology* **80**: 1354-1369.

Holdridge L. R. 1967. *Life zone ecology*. San José, Costa Rica: Tropical Science Center.

Hood G. M. 2004. *PopTools version 2.6.2*. http://www.cse.csiro.au/poptools.

Horvitz C. C. & Schemske D. W. 1995. Spatiotemporal variation in demographic transitions of a tropical understory herb: projection matrix analysis. *Ecological Monographs* **65**: 155-192.

Jongejans E. & De Kroon H. 2005. Space versus time variation in the population dynamics of three co-occurring perennial herbs. *Journal of Ecology* **93**: 681-692.

Larson R. J. 1992. Population dynamics of *Encyclia tampensis* in Florida. *Selbyana*, *13*, 50-56.

Lefkovitch L. P. 1965. The study of population growth in organisms grouped by stages. *Biometrics* **21**: 1-18.

Leimu R. & Lehtilä K. 2006. Effects of two types of herbivores on the population dynamics of a perennial herb. *Basic and Applied Ecology* **7**: 224–235.

Menges E. S. 1990. Population viability analysis for an endangered plant. *Conservation Biology* **4**: 52-62.

Menges E. S. 2000. Population viability analysis in plants: challenges and opportunities. *Trends in Ecology and Evolution* **15**: 51-56.

Mondragón D., Durán R., Ramírez I. & Valverde T. 2004. Temporal variation in the demography of the clonal epiphyte *Tillandsia brachycaulos* (Bromeliaceae) in the Yucatán Peninsula, Mexico. *Journal of Tropical Ecology* **20**: 189-200.

Pascarella J. B. & Horvitz C. C. 1998. Hurricane disturbance and the population dynamics of a tropical understory shrub: megamatrix elasticity analysis. *Ecology* **79**: 547-563.

Rzedowski J. 1986. *Vegetación de México*. (3rd ed.). Mexico: Editorial Limusa.

Schmidt G. & Zotz G. 2002. Inherently slow growth in two Caribbean epiphytic species: A demographic approach. *Journal of Vegetation Science* **13**: 527-534.

Silvertown J., Franco M., Pisanty I. & Mendoza A. 1993. Comparative plant demography – relative importance of life-cycle components to the finite rate of increase in woody and herbaceous perennials. *Journal of Ecology* **81**: 465-467.

Tabachnick B. G. & Fidell L. S. 1996. *Using multivariate statistics*. New York: Harper Collins.

Tremblay R. L. 1997. *Lepanthes caritensis*, an endangered orchid: No sex, no future? *Selbyana* **18**: 160-166.

Tremblay R. L. & Ackerman J. D. 2001. Gene flow and effective population size in *Lepanthes* (Orchidaceae): a case for genetic drift. *Biological Journal of the Linnean Society* **72**: 47-62.

Valverde T. & Silvertown J. 1997. A metapopulation model for *Primula vulgaris*, a temperate forest understorey herb. *Journal of Ecology* **85**: 193-210.

Valverde T. & Silvertown J. 1998. Variation in the demography of a woodland understorey herb (*Primula vulgaris*) along the forest regeneration cycle: projection matrix analysis. *Journal of Ecology* **86**: 545-562.

Warford N. 1992. *Erycina echinata. American Orchid Society Bulletin* **61**: 568-573.

Webster P. J., Holland G. J., Curry J. A. & Chang H.-R. 2005. Changes in tropical cyclone number, duration, and intensity in a warming environment. *Science* **309**: 1844-1846.

Wiegand T., Revilla E. & Moloney K. A. 2005. Effects of habitat loss and fragmentation on population dynamics. *Conservation Biology* **19**: 108-121.

Williams-Linera G. 1997. Phenology of deciduous and broadleaved-evergreen tree species in a Mexican tropical lower montane forest. *Global Ecology and Biogeography Letters* **6**: 115-127.

Winkler M. & Hietz P. 2001. Population structure of three epiphytic orchids (*Lycaste aromatica, Jacquiniella leucomelana,* and *J. teretifolia*) in a Mexican humid montane forest. *Selbyana* **22**: 27-33.

Winkler M., Hülber K. & Hietz P. 2005a. Effect of canopy position on germination and seedling survival of epiphytic bromeliads in a Mexican humid montane forest. *Annals of Botany* **95**: 1039-1047.

Winkler M., Hülber K., Mehltreter K., García-Franco J. & Hietz P. 2005b. Herbivory in epiphytic bromeliads, orchids and ferns in a Mexican montane forest. *Journal of Tropical Ecology* **21**: 147-154.

Wisdom M. J., Mills L. S. & Doak D. F. 2000. Life stage simulation analysis: estimating vital-rate effects on population growth for conservation. *Ecology* **81**: 628-641.

Zotz G. 1998. Demography of the epiphytic orchid, *Dimerandra emarginata. Journal of Tropical Ecology* **14**: 725-741.

Zotz G. 2005. Differences in vital demographic rates in three populations of the epiphytic bromeliad, *Werauhia sanguinolenta. Acta Oecologia* **28**: 306-312.

Zotz G. & Hietz P. 2001. The physiological ecology of vascular epiphytes: Current knowledge, open questions. *Journal of Experimental Botany* **52**: 2067-2078.

Zotz G., Laube S. & Schmidt G. 2005. Long-term population dynamics of the epiphytic bromeliad, *Werauhia sanguinolenta. Ecography* **28**: 806-814.

Zotz G. & Schmidt G. 2006. Population decline in the epiphytic orchid *Aspasia principissa. Biological Conservation* **129**: 82-90.

Zotz G. & Vollrath B. 2002. Substrate preferences of epiphytic bromeliads: An experimental approach. *Acta Oecologica* **23**: 99-102.

Die VDM Verlagsservicegesellschaft sucht für wissenschaftliche Verlage abgeschlossene und herausragende

Dissertationen, Habilitationen, Diplomarbeiten, Master Theses, Magisterarbeiten usw.

für die kostenlose Publikation als Fachbuch.

Sie verfügen über eine Arbeit, die hohen inhaltlichen und formalen Ansprüchen genügt, und haben Interesse an einer honorarvergüteten Publikation?

Dann senden Sie bitte erste Informationen über sich und Ihre Arbeit per Email an *info@vdm-vsg.de*.

Sie erhalten kurzfristig unser Feedback!

VDM Verlagsservicegesellschaft mbH
Dudweiler Landstr. 99
D - 66123 Saarbrücken

Telefon +49 681 3720 174
Fax   +49 681 3720 1749

www.vdm-vsg.de

Die VDM Verlagsservicegesellschaft mbH vertritt

Printed by Books on Demand GmbH, Norderstedt / Germany